ARKANSAS
Lake Port
LAKE PORT
AMERICAN CUT-OFF
CUT-OFF 1858
INDEFINITE
LAKE LEE
Lakeport
Avon
James
New Mexico Ch
Otter Lake
ISLAND NO 8
New Hope Ch
Pilgrims Rest Ch
North Star
ISLAND NO 87
WORTHINGTON TOWHEAD
HOWARD ISLAND
Longwood
R-I-V-E-R
BEND
KENTUCKY
FANNY BEND
WORTHINGTON POINT
ISLAND NO 88
WORTHINGTON CUT-OFF 1933
WASHINGTON COUNTY
Chatham
Pleasant Stay
Byrne City
Grand Lake
GRAND LAKE
Willow Lake
CUT-OFF 1796-1817
MISSISSIPPI
LAKE WASHINGTON
Marathon
Daniels Chapel

MISSISSIPPI FLOODS

MISSISSIPPI

DESIGNING

Anuradha Mathur & Dilip da Cunha

FLOODS
A SHIFTING LANDSCAPE

YALE UNIVERSITY PRESS NEW HAVEN AND LONDON

for Tara

Published with assistance from the Graham Foundation for Advanced Studies in the Fine Arts and from the University of Pennsylvania Research Foundation.

Design
Henk van Assen, NY
Typefaces
Thesis Sans, Din Mittelschrift
Printing
C & C offset Printing Co.
Printed in Hong Kong

Library of Congress Cataloging-in-Publication Data
Mathur, Anuradha.
Mississippi floods : designing a shifting landscape / Anuradha Mathur and Dilip da Cunha
p. cm.

Exhibition catalog. Includes bibliographical references and index
ISBN 0-300-08430-7 (cloth)
1. Mathur, Anuradha—Exhibitions. 2. Cunha, Dilip da—Exhibitions. 3. Mississippi River Valley—In art—Exhibitions. 4. Landscape assessment—Mississippi River Valley. I. Cunha, Dilip da. II. Title

N6537.M3943 A4 2001
917.7'02—dc21 00-011101

A catalog record for this book is available from the British Library.
The paper in this book meets the guidelines for permanence and durability of the Committee on Production Guidelines for Book Longevity of the Council on Library Resources.

10 9 8 7 6 5 4 3 2 1

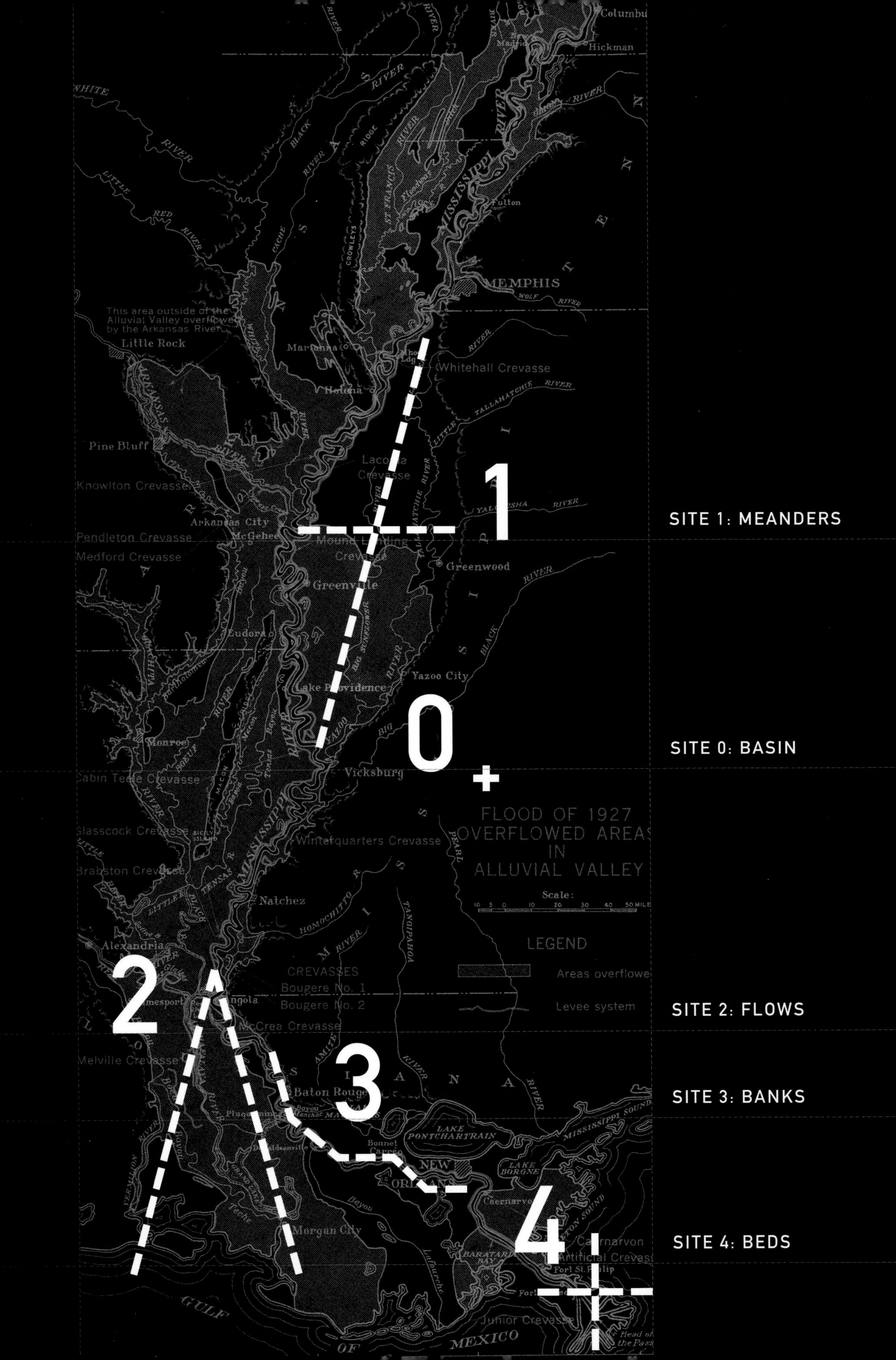
1
0
2
3
4
SITE 1: MEANDERS
SITE 0: BASIN
SITE 2: FLOWS
SITE 3: BANKS
SITE 4: BEDS
FLOOD OF 1927
OVERFLOWED AREAS
IN
ALLUVIAL VALLEY
Scale:
LEGEND
Areas overflowed
Levee system
CREVASSES
Bougere No. 1
Bougere No. 2
McCrea Crevasse
This area outside of the
Alluvial Valley overflowed
by the Arkansas River
Little Rock
Pine Bluff
Knowlton Crevasse
Arkansas City
McGehee
Pendleton Crevasse
Medford Crevasse
Greenville
Greenwood
Eudora
Lake Providence
Yazoo City
Monroe
Vicksburg
Cabin Teele Crevasse
Glasscock Crevasse
Winterquarters Crevasse
Brabston Crevasse
Natchez
Alexandria
Melville Crevasse
Baton Rouge
Bonnet Carre
LAKE PONTCHARTRAIN
NEW ORLEANS
LAKE BORGNE
Caernarvon
Caernarvon Artificial Crevasse
Fort St. Philip
Junior Crevasse
Morgan City
MEMPHIS
Whitehall Crevasse
Marianna
Helena
Hickman
Fulton
Mound Landing Crevasse
Laconia Crevasse
GULF OF MEXICO

PREFACE

In 1993 the media displayed vivid images of a Great Flood spreading across the American Midwest, unsettling lives and property on more than ten thousand square miles of land. The Mississippi River was defying its captors in what the United States Geological Survey deemed the costliest and most devastating flood in American history. But even as America's River was demonstrating a material elusiveness, it was impossible not to notice that it was being captured by aerial photographs, satellite pictures, flood extent maps, height markers (it was flowing forty-six feet above normal, three feet higher than had ever been recorded), probability measures (its spread was deemed a five-hundred-year flood), and news reports. We were intrigued. Behind the scenes of breaking levees, moving houses, desperate sandbagging, and anguished faces was a "Benchmark Mississippi" to which the extensive data and commentary on the Great Flood referred. The mighty Mississippi, trying as hard as it was, seemed unable to wash away this taken-for-granted firm ground. It was perhaps the only ray of hope in this disaster, assuring inhabitants that even if the Mississippi was successful in overtopping its levees this time, it would eventually be subdued and, using the measures of this flood, its next escape would be made a lot more difficult.

Even while the Midwest was still a shimmering, muddy brown sea, however, the measures were proving divisive. In a situation of perceived winners and losers, the questions being asked in the search for cause and blame were polarized, reflecting nothing of the muddiness of the scene.[1] Were we witnessing a "natural" disaster exacerbated by the wettest June and July since 1895 (twelve major storms with rainfalls of six to twelve inches) or a "cultural" tragedy caused by the constricted flow of this continental drain? Was this the fault of a river out of control or of settlement in the floodplain? The responses to these questions, predictably perhaps, tended to move in opposing directions. From one side we heard calls for more control of the river; from the other, for the withdrawal of settlement from the floodplains. The two sides were reinforced by the dual reading of flood as a giver and a taker of life: facilitating emergent life, on one hand, destroying habitats on the other. But it was not the differences between the extremes that drew our attention. It was rather their common ground. While this ground held the spirit of compromise, it told us that even as the Mississippi was erasing property lines and dissolving boundaries of all kinds it was not shaking the distinctions by which this landscape has been looked upon and inhabited for the past three centuries: river and settlement, nature and culture, water and land.

We were provoked by this "landscape of flood," a landscape comprising material and ideological constructions that hold the mighty Mississippi captive even after it breaks through the most elaborate and calculated control system in the world, designed and maintained by the U.S. Army Corps of Engineers. These landscape designers and indeed the public at large would have us believe that in preparing for the next flood, we must further tame the Mississippi, surrender the floodplain, or find a more effective middle way. Looking at the terrible toll of flood in the most technologically advanced nation on earth, we were troubled by these choices. They were underpinned by the view of the Mississippi as an object, an object that can be controlled or released, an object that can be subject to categories.

We resolved to set out on a journey, naively perhaps, to re-engage the Mississippi, not as an object, but as a dynamic, living phenomenon that asserts its own dimensions. This book is an account of that journey through the Mississippi mud. It was truly the inhabiting of a shifting terrain. Each day brought new revelations, introduced us to new ways of seeing things, ways that constantly changed the course of our travels. We stumbled upon sites and followed leads, collecting material that appears in the illustrations that follow: map-prints (screen prints in which we layered and erased information), photo-transects (photomontages in which we brought together the maps and horizons of our journey), historical maps, line drawings, and paintings. Many of these illustrations have been shown in an exhibition, *Mississippi Horizons: Mapping a Shifting Terrain,* that began at the University of Pennsylvania in the fall of 1998 and has since traveled to other venues in the United States and Europe. Although the exhibition and the book each stand on their own, they benefit from reciprocity.

We have a number of people and organizations to thank for a journey that was as exciting and rewarding as it was unpredictable—and that continues to be incomplete. It was initiated in 1996 with a grant from the Research Foundation of the University of Pennsylvania and a first-hand introduction to the world of Mississippi hydraulics in a trip to Vicksburg, the Old River Control Structures at Simmesport, the Atchafalaya Swamp, and the River Corridor from Baton Rouge to New Orleans. At the Mississippi River Commission in Vicksburg the late Michael Robinson gave us access to a room full of maps, publications, and working documents. John Brooks indulged our vague questions on the engineering of the Mississippi, undoubtedly puzzling over our purpose. We visited the Mississippi Basin Model at Clinton and owe it to Dinah McComas of the Waterworks Experiment Station in Vicksburg for helping us find the 1940s drawings of the layout and construction of this miniature feat of river hydraulics. She evidently valued them as much as we did. At the headquarters of the U.S. Army Corps of Engineers in New Orleans, James Addison, Nancy Mayberry, and Tom Hasenboehler allowed us access to their slide collection and working drawings of the Control Structures at Simmesport. Following a lead from John McPhee's insightful book *Control of Nature*, we discovered the indefatigable Professor Charles Fryling at Louisiana State University in Baton Rouge. His passion for the Atchafalaya Swamp was an inspiration, and the daylong canoe excursion that he took us on into this serene and wondrous place with its ever-changing rhythms is unforgettable.

We returned to the Mississippi in the summer of 1997 in search of the Greenville Bends, the Delta blues, Stack Island, and the Cajun Triangle among numerous other phenomena that we had sighted in the many maps, images, and writings we had perused over the months. We were able to find a wealth of information at catfish farms and factories, gas stations, revetment fields, experimental farms, churches, motels, bed and breakfasts, hunting clubs, crossings, even the Delta Regional Medical Center, where one of us was admitted with dehydration. We had our share of good fortune, too. Crossing a deserted street in downtown Clarksdale one hot Sunday afternoon we were approached by blues musician Arthneice Jones, leader of the "Stone Gas Band." He treated us to a commentary on his own music and the state of the blues in the delta. Standing in the middle of the road, this man, a concrete-layer by trade, put our journey in perspective by reciting one of his lyrics: "I did not search for the blues, the blues found me." Lyle Pringle of the Delta Research Center provided us with information about the agricultural rhythms and practices of the delta. Attorney Robert Bailess in Vicksburg generously gave us access to the map sequence with which he successfully argued the continued place of Stack Island in the State of Mississippi despite the island's "migration" across the river into the State of Louisiana.

All along our travels we yearned to be on the other side of the levee, on one of the greatest navigational systems in the world, a system that we were able to only glimpse in the occasional bridge crossing and trespass on the levee. Our chance eventually came with the generosity of OrGulf, Ltd., based in Paducah. We boarded the towboat MV *Jim Ludwig* in November 1997 as it passed Memphis on its way to New Orleans. The eight-person crew with Captain Riley Powell at the helm made us feel at home for six memorable days. The boat was a remarkable classroom and Captain Powell, like Mark Twain's Mr. Bixby, a great teacher. For those few days we felt Twain's "permanent ambition" to be a riverboat pilot. A corner of the wheelhouse, fifty feet above the water, became our moving studio, the most stimulating one we have ever occupied. It was certainly the best place to feel the might of the Mississippi.

We made another trip to New Orleans in August 1998 to fly over Pilottown, an amphibious settlement that is the gate to the continent, in the mouth of the Mississippi. Pilot Claude ("Skip") Blanchard of Air Router, New Orleans, flew us over this emergent terrain. It is a wonder how this fluid world below us was ever mapped before the advent of aviation.

Back home on the drawing board in Philadelphia we struggled to accommodate the shifting landscape of our travels, to make sense of the multitude of images, maps, and documents that we had gathered. The journey continued in the printmaking studios of the Department of Fine Arts at the University of Pennsylvania. Hitoshi Nakazato allowed us the run of the studios in the summer of 1998. But if it had not been for the inspiration and guidance of Chad Andrews, we would not have had the courage and wherewithal to negotiate the temperamental environment of the studios and the often elusive processes of screen printing.

We were supported at various stages of this work by James Corner, Robert Giegengack, Herbert Gottfried, Gary Hack, John Dixon Hunt, Ian McHarg, Michael Meister, Elizabeth Meyer, Mohsen Mostafavi, Adele Santos, Robert Slutzky, and Anne Whiston Spirn. We thank them as well as Ramsey Silberberg, Charles Neer, Wookju Jeong, and Adam Greenspan, who helped our travels enormously with their research and labor. We appreciate the encouragement of family and friends, in particular Rajmohan Shetty, Alok Mathur, Vincent da Cunha, Tom Leader, Diane Karp, Joan Pierpoline, Venkatesh Babu, Kavita Khanna, and Francis and Vatsala da Cunha. At Yale University Press we are grateful to Judy Metro for taking on this project, Mary Mayer for seeing it through, Phillip King for his editorial finesse, and Henk van Assen for his design insight. Publication of this book, in particular its color images, was supported by a grant from the Graham Foundation for Advanced Studies in the Fine Arts and the Research Foundation of the University of Pennsylvania.

Finally, our daughter, Tara. Born in the year 2000, she is already a fellow traveler of shifting landscapes. We dedicate this book to her.

INTRODUCTION

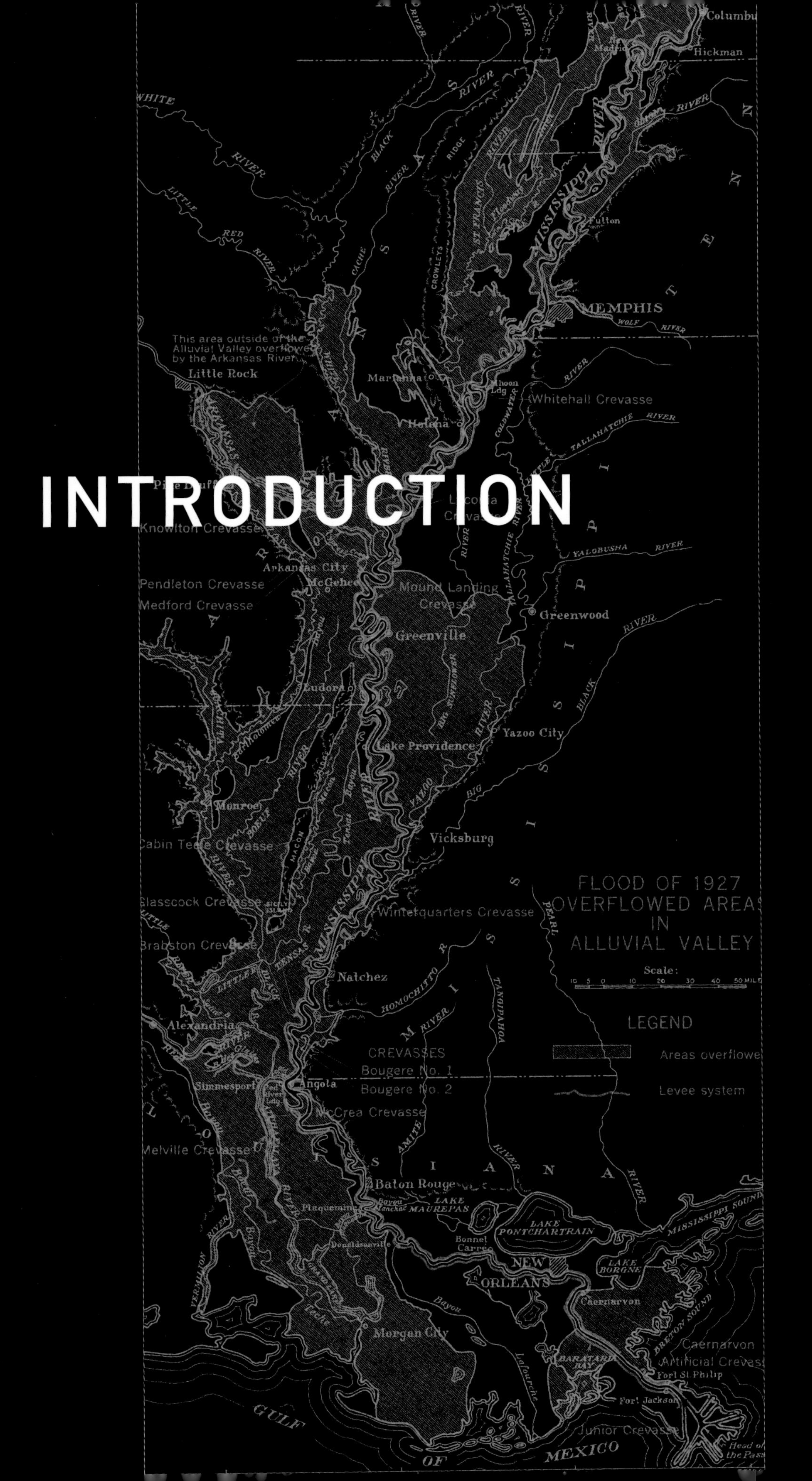

MISSISSIPPI HORIZONS

A levee break in the Great Flood of 1927 that submerged 23,000 square miles from Cairo, Illinois, to the Gulf of Mexico, reportedly destroying more than 160,000 homes and killing between 250 and 500 people [U.S. Army Corps of Engineers]

The Mississippi floods. It floods despite heroic efforts to confine its course. These floods recur—as they have most recently in 1993, 1995, and 1998—with billions of dollars in measurable destruction and immeasurable distress. They are a concern not only for the American heartland but also for the many nations that aspire to the flood control measures employed on the Mississippi. The Mississippi after all nourishes and drains 41 percent of the world's most powerful—and wealthy—nation. Yet the war that it has declared on the Mississippi is far from won or, for that matter, lost. It has merely become an everyday practice, embodying conflict in a unique landscape of levees, cutoffs, revetments, overflow channels, floodways, locks, gates, and jetties.

This landscape of conflict, the result of efforts to prevent floods while exploiting the Mississippi's unrivaled navigational potential and, importantly today, valuing its ecological role, is itself the subject of much dispute. The sense of permanence, security, and prosperity that designed interventions have promised is being increasingly questioned in Congress and by individuals, in academia and the popular press. While some continue to believe that the Mississippi can be harnessed and controlled, others, notably radical environmentalists, advocate that it be released into some natural state. Proponents of these views, and of many others, take their positions with regard to a landscape already constructed, its horizons defined, but meanwhile little attention is paid to the representations that play a significant role in constructing the Mississippi that is the subject of these views. These representations include maps, hydrographs, cross sections, working drawings, and models used by professionals in the process of designing this landscape, but also, in a more popular vein, photographs, media reports, paintings, and folklore that have contributed to the reception and inhabitation of this designed landscape. These representations reveal a shared idea of the Mississippi, no matter how diverse the views, a Mississippi already captured by a larger cultural imagination before the war on the ground is fought. They hold the promise of a discourse not merely on the divisive views of the Mississippi but on the imaginative content of its construction.

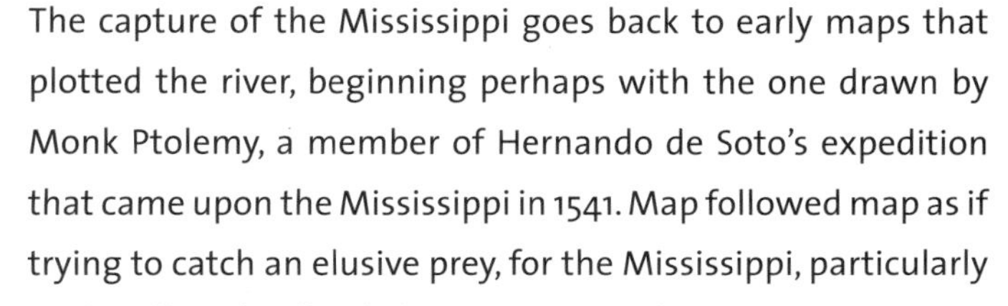

The capture of the Mississippi goes back to early maps that plotted the river, beginning perhaps with the one drawn by Monk Ptolemy, a member of Hernando de Soto's expedition that came upon the Mississippi in 1541. Map followed map as if trying to catch an elusive prey, for the Mississippi, particularly in the alluvial valley below St. Louis—the Lower Mississippi—has changed course a number of times and even swelled to occupy nearly all of the valley on a couple of occasions in living memory. It was a different phenomenon to Native Americans who inhabited a land that diminished and expanded, with the rise and fall of the "Father of Waters," rather than drained. But it was not until the professional surveys beginning in 1820, commissioned by the United States Congress for the purposes of navigation and performed by the Corps of Topographical Engineers and the Army Corps of Engineers, that the Mississippi began to take a definitive form with regard to shape, depth, slope, discharge, material, and hydraulics. It is a form that made Congress conceive granting the Army Corps of Engineers the authority, following the Civil War and several floods, to "keep the Mississippi in its path and out of politics." Even if theories of river dynamics and strategies of river control have changed, the course of the Mississippi has hardly strayed from the form determined for it by these surveys: a river whose waters can in some way be contained or, if pressure mounts, released. When the army's engineers declare victory as they did on one occasion—"we harnessed it, straightened it, regularized it, shackled it"—they do so over an opponent that they have already captured in their surveys. When they are forced to give in, for example by the Great Flood of 1993, they submit to an adversary that they have imaged. [1]

A levee break in the Great Flood of 1993 that inundated 15,600 square miles of the Midwest, destroying 55,000 homes and killing 50 people [Chris Stewart, Blackstar]

F. E. Palmer, *High Water in the Mississippi*, 1868 (lithograph, 18" x 28") [Knox College Archives, Galesburg, Illinois]

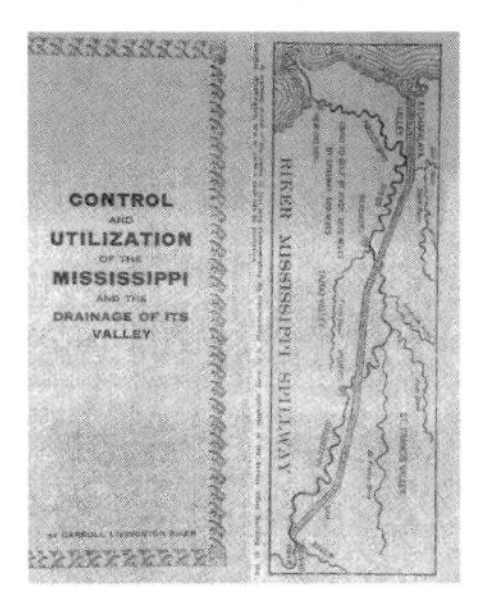

Cover of the pamphlet advertising Captain Carroll Riker's proposal to the U.S. Congress in 1929 for a 500-mile-long spillway from Cairo to the Gulf, bypassing 1,070 miles of the Mississippi River and carrying more than twice the volume of the 1927 flood

No less significant than the surveys of professionals authorized to care for this landscape are the more popular images of the Mississippi as a river, particularly a river in flood when it is "a savage clawing thing right at the top of the levee and sounding at night like the snarl of a beast." The awe-inspiring, and often heartrending, images of flood have been significant in the public eye from at least the time when Garcilaso de la Vega, another member of De Soto's expedition, described a river "that rose to the top of the cliffs . . . to overflow the meadows in an immense flood." Three hundred and thirty years later Samuel Clemens, who was born in Hannibal on the Mississippi and took his pen name, Mark Twain, from the boatman's measure of its depth (meaning two fathoms or twelve feet), provided the view from the river. In *Life on the Mississippi*, which recounts his education and adventures as a riverboat pilot, he describes the "watery solitude" of a "lawless stream" in flood: "majestic, unchanging sameness of serenity, repose, tranquillity, lethargy, vacancy—symbol of eternity, realization of the heaven pictured by priest and prophet, and longed for by the good and thoughtless!" [2]

The violence of a levee break and the solitude of the aftermath: two horizons of flood
[Top: U.S. Army Corps of Engineers. Bottom: National Geographic Society, Image Collection]

The lawless Mississippi must have appeared differently to the black slaves who cleared, drained, protected, and cultivated its banks and remained second-class citizens until well past the middle of the twentieth century. Their melancholic cries in flood are heard in the Delta blues, a music that evolved in the Yazoo Delta immediately south of Memphis. Charlie Patton in his classic "High Water Everywhere" sings of the ravages of the 1927 flood and the hilly country from which he was barred. It was a time when aerial photography was extending the horizon of flood. Dramatic images of inundation and rescue made headlines in far-reaching newspaper and magazine reports. In times of flood today, we are inundated with television documentaries and satellite pictures instantly beamed back from hundreds of miles in space. Surely these images of tragedy and wonder, with varying means and extents of dissemination, influence public policy, the popular press, and emotions regarding the "Father of Rivers," "Old Devil River," "Old Big Strong," "Big Muddy," the "Great Common Sewer."

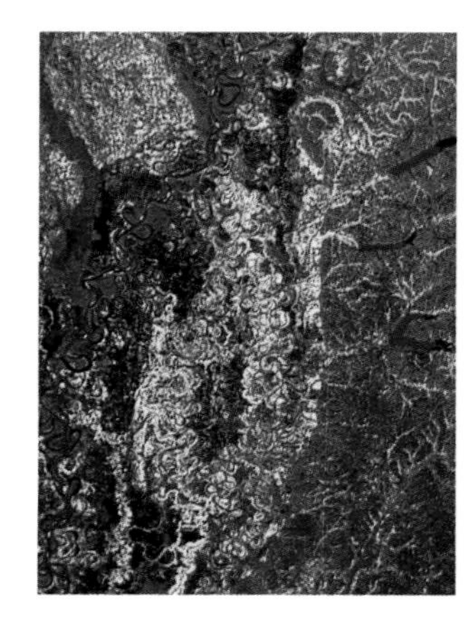

Landsat digital mosaic of the Memphis–Vicksburg region of the Mississippi alluvial valley
[Veridian Erim International]

A majority of these names that acknowledge the river's power, danger, and burden refer to the Lower Mississippi, which begins at Cairo, Illinois, where it collects the flow of its largest contributor of water and traffic, the Ohio River, and shortly after it is joined by its longest and most sedimented branch, the Missouri. At this point, as Willard Price describes it, the "modest river hiding within its walls becomes a brazen exhibitionist riding on top of the world. . . . The river behind us is a dug-out river. It has made a trench for itself and the rock walls of the trench rise sometimes three hundred feet high. The river ahead of us had an entirely different idea. It rides on top of the land, somewhat like an aqueduct. You look down upon the upper river from the precipices that contain it. You look up to the lower river from lands that have to be protected from it by levees."[3]

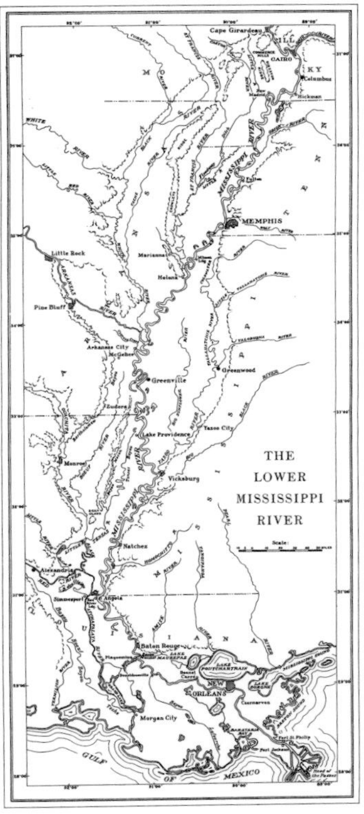

Map of the Lower Mississippi River, from D. O. Elliott, *The Improvement of the Lower Mississippi River for Flood Control and Navigation*, 1932

Both the upper and lower rivers are powerful agents in the landscape. But if the upper river has dug itself in, the Lower Mississippi has roamed the surface of the valley from Cairo to the Gulf of Mexico like a monarch conjuring up mythical figures of the River King, but also making it difficult to separate the river from the land it makes. The change in agency was noticeable to Captain Frederick Marryat. Writing in 1839, he observed: "The waters of the present upper Mississippi are clear and beautiful; it is a swift but not an angry stream, full of beauty and freshness," whereas the Lower Mississippi was an "impetuous, discolored, devastating current . . . constantly sweeping down forests of trees in its wild course, overflowing, inundating, and destroying, and exciting awe and fear." It is this force that has enabled the Lower Mississippi to lay the alluvium of this valley, thirty to ninety miles wide and six hundred miles long, over the past one million years, pushing back a sea. Even today its reach extends as it spews, in the description of one commentator, twenty-five thousand railroad cars' worth of mud into the Gulf of Mexico every day. It is a sight as visible today as it was to Captain Marryat: "There are no pleasing associations connected with this great common sewer of the western America which pours out its mud into the Mexican Gulf polluting the clear blue sea for many miles beyond its mouth."[4]

It is the landscape of the River King that is the focus of this book. It may not shift as regularly and visibly as it once did. But looking beyond the confinements of levees, locks, gates, and so on, to the representations employed in their design—maps, hydrographs, photographs—one clearly discerns a shifting landscape. The public rarely gets to experience this shifting landscape, except perhaps in flood, when wonder at its magnificence is tinged with the tragedy of its devastation.

A PANORAMA

This book is an attempt to bring into public discourse the images instrumental in the design of the Lower Mississippi, thereby broadening appreciation for the life of a wondrous landscape. In this it carries forward the tradition of the panorama, a popular mode of entertainment by which the Mississippi landscape was brought to public attention in the nineteenth century.

Often described as the precursor to the modern newsreel and documentary, panoramas were remarkable feats of painting on canvas. They were distinctive for their size but also for their ambition to present landscape as a continuous and inexhaustible phenomenon. In the moving panoramas, a continuum of changing horizons passed in front of spectators. John Banvard of St. Louis, the first to engage an American and European audience with his *Mississippi Panorama*, announced that it was "Painted on Three Miles of canvas, exhibiting a View of the Country 1200 Miles in Length, extending from the Mouth of the Missouri River to the City of New Orleans, being by far the Largest Picture to be ever executed by Man." The handbill of another panorama, by John J. Egan, based on the archeological explorations of Prof. M. W. Dickeson, claimed to put the "Monumental Grandeur of the Mississippi Valley" on fifteen thousand feet of canvas. The press applauded these works for their realism and accuracy. Most, like Henry Lewis, did indeed aspire to represent "a truthful view of the river and all the principal objects on its shore the whole distance."[5]

The machinery for John Banvard's Moving Panorama, from *Scientific American*, December 16, 1848

The map-prints, photomontages, line drawings, paintings, and essays featured in this book likewise seek to portray the grandeur of the Lower Mississippi as an inexhaustible phenomenon. But their collective portrayal of the Lower Mississippi differs from these great works of the nineteenth century in one important respect. The landscape that they present is more than scenic. It is "working." This is not to diminish the beauty of the lower river, although beauty can hardly be discerned in accounts of this river, particularly those of travelers and novelists. Phrases like "melancholy dreariness," "dismal vastness," "eternal gloom," "fearful wildness," "filthy waters," "lawless stream," and "wretched place" are common in descriptions of the river scene. Many have remarked how considerably less sightly the Lower Mississippi is compared with any other river, particularly the Upper Mississippi. A visitor in 1837 went as far as to say: "It is not like most rivers, beautiful to the sight, bestowing fertility in its course; not one that the eye loves to dwell upon as it sweeps along, nor can you wander on its banks, or

Handbill advertising the Egan-Dickeson Panorama [University of Pennsylvania Museum, neg. no. S4.143336]

trust yourself without danger to its stream. It is a furious, rapid, desolating torrent, loaded with alluvial soil." Sightliness, however, is a secondary concern. The landscape we portray is not something to look at as much as a process, much of it taking place when we are not looking and, furthermore, where we cannot easily look.[6]

Some early-nineteenth-century travelers to the Mississippi sensed a process: "it was not till after I had visited the same and similar spots a dozen times, that I came to a right comprehension of the grandeur of the scene." Others found that the landscape required use of the imagination. Charles Augustus Murray, coming upon the Father of Waters via the Ohio, overcame initial disappointment at the muddy sight: "It is only when you ascend the mighty current for fifty or a hundred miles, and use the eye of the imagination as well as that of nature, that you begin to understand all his might and majesty. You see him fertilizing a boundless valley . . . here carrying away large masses of soil with all their growth, and there forming islands, destined, at some future period, to be the residence of man."[7]

CONSTRUCTING THE MISSISSIPPI

Top left:
Building levees
[U.S. Army Corps of Engineers]

Bottom left:
Laying revetments
[U.S. Army Corps of Engineers]

Top right:
Dredging the channel
[U.S. Army Corps of Engineers]

Bottom right:
Dynamiting a levee
[Clifton Adams, National Geographic Society Image Collection]

The panorama of the Lower Mississippi that we are trying to paint here recognizes a river that is itself boundless, carrying the diversity of a basin, its soils, seasons, peoples, histories, technologies, connections, and conflicts, reason enough for Mark Twain to call it the "body of the nation." Its agency, however, in moving masses of soil and forming islands is mixed in no small measure with everyday human practices of cultivating, inhabiting, dredging, constructing, maintaining, and, not the least, negotiating. And more than this human agency that pushes the eye of imagination beyond conventional ways of seeing, the Mississippi carries a design agency—the constructions conceived on the professional's drawing board, where images of the landscape in the process of design play a determining role in the outcome. To portray a working landscape, we bring forth from behind the scenes the body of a nation, the everyday human practices that contribute to the construction of the Lower Mississippi, and, more specifically, the images that play a role in the process of designing this landscape.[8]

The objective of displaying a working rather than scenic landscape is to cultivate a critical public. In this regard, this work is inspired by an increasing number of scholars and activists who in a recent surge of interest and concern have sought to reveal representations, particularly maps, as powerful ideological, colonial, cultural, and even fictional instruments in the service of power.[9] The premise is that images are projective rather than descriptive, and that their truth lies not merely in what they portray but also in what they leave out. But, more than critique, this panorama responds to a constructive and communicative urgency. It is directed toward finding creative ways to engage a dynamic landscape—a terrain that is never still, always shifting. It calls the attention of designers and the public to a terrain whose shifts are visible in the images that have led to the designed landscape but hardly evident in the landscape itself, until, that is, the Mississippi floods.

TRAVELING THE MISSISSIPPI

In 1993, the Upper Mississippi system above Cairo flooded the Midwest devastatingly. It was as if the muddy waters of the north were unable to find or squeeze through the outlet designed for them by the Corps of Engineers—the Lower Mississippi that winds a thousand miles to the Gulf. While the flood renewed questions about the battle between man and river, perhaps the most important point it made was that the design of the Mississippi, which began with the first levee construction in the 1710s to protect the birth of New Orleans, is evidently still on the drawing board. Whether the engineers of the Corps build levees up to each and every source in the basin, emphasizing the Mississippi's status as a continental drain, or whether they release the Mississippi to run free in part or whole, they do it by design. The questions that the 1993 flood raised for us were—what is the Mississippi that sits on the drawing board, what are the measures that have stilled this "snarling beast" enough to subject it to design? Are there other measures, more accommodating of its shifts?

The abandoned Mississippi Basin Model, Clinton, Mississippi, 1997

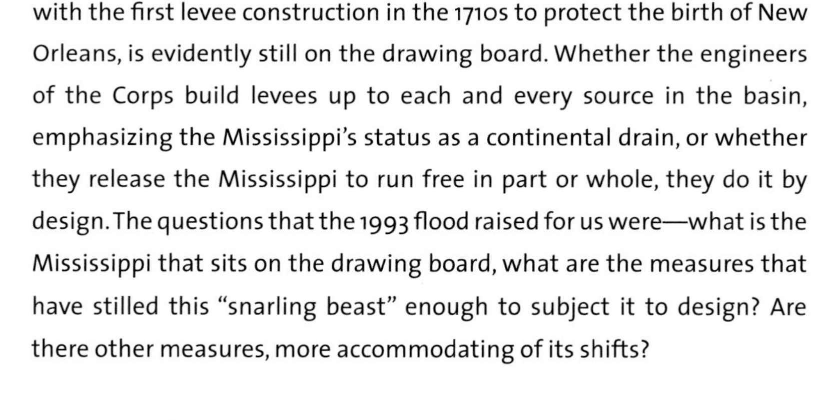

In 1996 we began a journey across the Lower Mississippi. It took us through documents—maps, surveys, engineering drawings, newspaper reports, books—as well as a varied material terrain—sites with rich and layered phenomena and the foci of repeated surveys and conflict. The recounting of that journey is this panorama of a working landscape.

The panorama begins not with a place on a river but with the Mississippi Basin Model, a working model of the river located forty miles away from it, at Clinton, Mississippi. Built in the 1940s by the Army Corps of Engineers, the model marks a culmination of an era of maps and surveys, and a starting point in the design of significant control structures on the

Mississippi. We call this cumulative point of departure Site 0. From here we travel four terrains, each the focus of repeated surveys and conflicts, each a force in its own right in the construction of the Mississippi landscape, each elusive. They are *meanders*, *flows*, *banks*, and *beds*. We call them Sites 1, 2, 3, and 4.

The elusiveness of these terrains has not stopped the Corps from pursuing their definition. The efforts of the Corps have resulted in interventions that match the Mississippi in magnificence. These interventions have an impact not only on the river but on larger lands of the Mississippi's making. They also bring life to these lands, forcing them into the inexhaustible gray zone between man and river that marks the Lower Mississippi landscape.

FOUR "CONSTRUCTION SITES" ON THE LOWER MISSISSIPPI

Top to bottom:

The Greenville cutoffs near Greenville, Mississippi

The Old River Control Structures at Simmesport, Louisiana

The River Road from Baton Rouge to New Orleans

The Passes at the mouth of the Mississippi River, Louisiana

[All photos: U.S. Army Corps of Engineers]

MEANDERS

We were led to Site 1 through the engineered cutoffs of the Greenville Bends, about ninety miles northwest of the Mississippi Basin Model. Not only do the cutoffs figure repeatedly as a textbook case in river engineering, but the bends themselves are legendary in accounts of the Mississippi. They are a series of S-shaped turns in the river separated by tantalizingly narrow necks of land. The Corps cut through the necks and many others up and down the stretch of the river between Memphis and Vicksburg in the 1930s and '40s. In the process of this straightening they stilled the meandering power of the Mississippi, a power with which the river has laid the soil of the valley and left its imprint on the valley's surface. Nowhere is this imprint more clearly portrayed in satellite pictures than in the neighboring Yazoo Delta, the last of the bottomlands to be brought under the plantation regime in the mid-1800s. The delta opened a world, not only of former paths of the Mississippi, but also of the resistance and rhythms of the Delta blues, a music with roots in the hollers of slaves who worked the cotton fields and built the railroads and levees. Noted for its elusiveness, blues is the voice of this landscape cleared and drained for plantations. In the delta the engineered curves of the Corps reside uneasily with the meanders of the blues.

FLOWS

Site 2 was opened to us by the controversial accounts of the control structures at Simmesport, Louisiana, a site popularized by John McPhee in *Control of Nature* as the most direct confrontation between the Mississippi River and the U.S. government. Here the Mississippi for decades has been suspected of wanting to change its course, to go down the Atchafalaya River and leave the settlements on the New Orleans–Baton Rouge corridor without a river. To prevent this takeover, the waters of the Mississippi (and of the Red River, which meets it here) are vigilantly monitored and divided by the structures at Simmesport. In the process of keeping the Atchafalaya from capturing the Mississippi, however, the Corps of Engineers has also inherited the flow of eight hundred thousand acres of labyrinthian bayous and swamps of the Atchafalaya basin, a place that environmentalists despite the massive human agency involved would like to see declared a natural reserve. The basin is not easy to manage. This living, breathing terrain is home not only to the bayous that are famous for defying any rules of movement, and the subtle resistance of the Cajuns who live off them, but also to a multiplicity of conflicting interests, each desiring a different flow.

BANKS

Site 3 originated with the levees that appeared in maps and accounts in the early eighteenth century and steadily extended to define a river corridor. Although these feats of river engineering today line almost the entire length of the Lower Mississippi, they are most prominent both on the ground and in politics along the corridor between Baton Rouge and New Orleans. Here, more than anywhere else, they channel the river, hastening silt and sediments to the Gulf. These manicured earthen embankments have risen in three centuries from four feet to over forty feet. They keep the Mississippi from overflowing, a crossing of water that once raised and enriched its banks with the deposition of land that has so contentiously been cultivated first by a plantation culture and today by a booming petrochemical and transportation industry. The segregations of the plantations are still visible on either side of the river in antebellum homes, oak alleys, sugarcane fields, and poor shacks. They reside uneasily amid the integrative infrastructure of the modern industries: bridges, wires, pipelines, and loading belts that operate around the clock to the rhythms of a global corporate culture.

BEDS

We entered Site 4 through the legendary design conflict over the opening of the passes at the mouth of the Mississippi, one hundred miles below New Orleans. The battle between the civil engineer James Buchanan Eads, designer of the first steel bridge at St. Louis, and the Corps of Engineers is well documented. It was a fight over the choice between the strategic construction of jetties to urge a Mississippi laden with the sediment of a continent to dredge its own bed and the daily dredging of the passes sanctioned by the Corps. Eads won the battle in the late 1800s and opened the continent to dependable trade. Since then jetties have taken control of the river's beds, not just in the passes but at many places upriver too. Even as the jetties at the mouth ease the way for the tons of mud and vessels over the continental shelf, they also trap soil and make new land. In this desolate and forbidding place—the Birdfoot Delta—where surfaces are constantly being remade, Pilottown is the most recent of a number of amphibious settlements (others before it having disappeared without a trace) that facilitate the exchange between ocean and river pilots. Here, on the threshold of its release, one sees the emergent power of the Lower Mississippi.

The panorama of the Lower Mississippi presented in the following pages travels these sites through short essays and images—map-prints, photo-transects, and paintings. The photo-transects and paintings record the horizons of our travel—walking on the basin model, driving across the Yazoo Delta, canoeing in the Atchafalaya, boating down the River Corridor, and flying over Head of the Passes. The map-prints, on the other hand, are generative and explorative documents. They are hand-pulled screen prints that use a range of representations—found maps, line drawings, notations, media images, text, and data—to draw out issues and measures of Sites 1, 2, 3, and 4. The medium of screen printing afforded a unique way to sediment and erase, layer and juxtapose, reveal and conceal the wealth of information that confronted us as well as to structure relationships and appreciate serendipitous occurrences. It makes the map-prints constructions in addition to presentations of the landscape that we traversed. The brief essays that accompany the map-prints extend this construction of the Lower Mississippi.

SITE 0 BASIN

BASIN MODEL / MISSISSIPPI BASIN

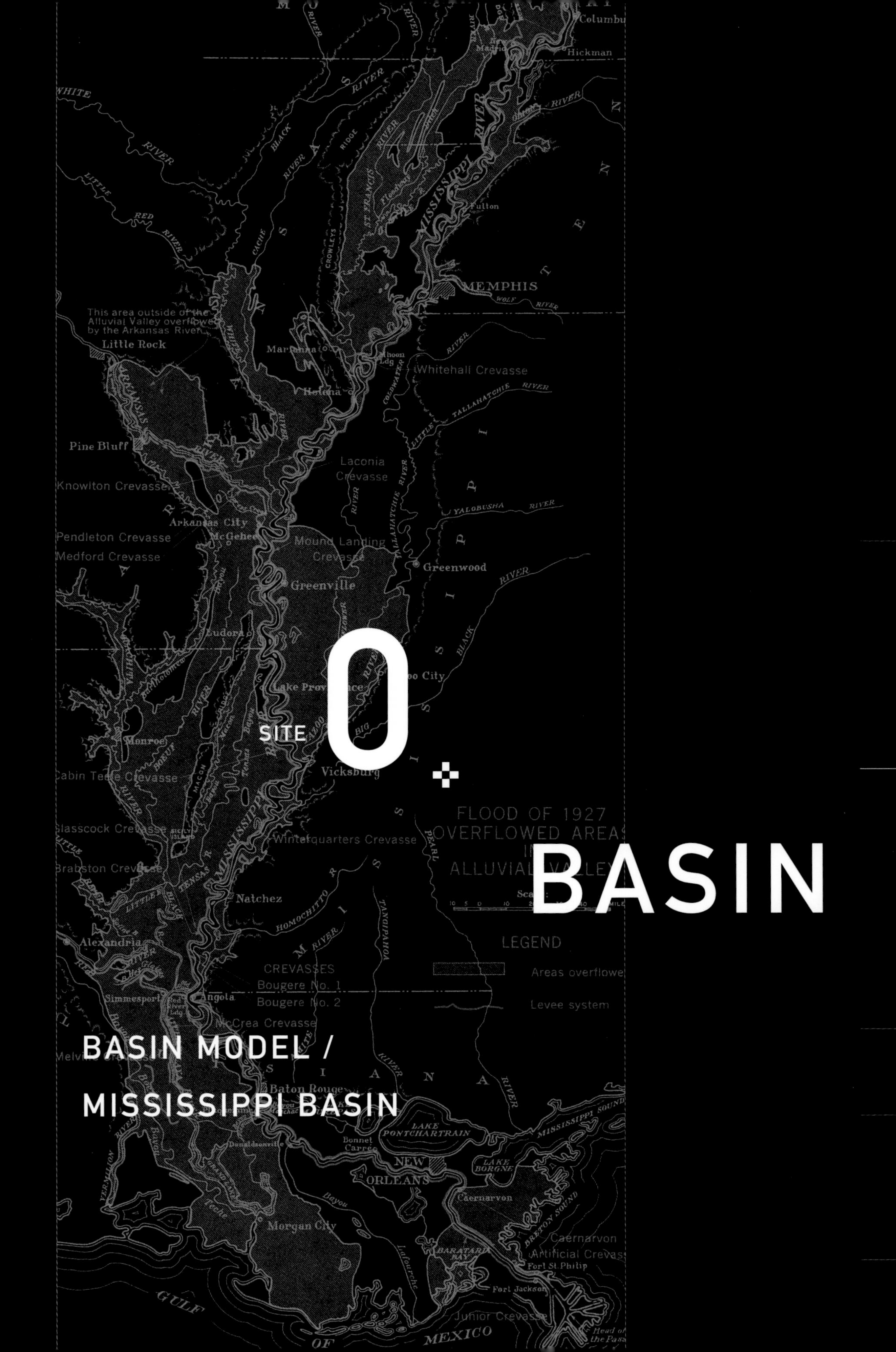

THE BASIN MODEL

In the township of Clinton, Mississippi, on the outskirts of Jackson, forty miles due east of Vicksburg on the Mississippi River, lies abandoned the world's most ambitious working model—the Mississippi Basin Model. It covers forty acres of land, leveled and molded into a giant relief map. A vertical foot on the model equals one hundred feet of the Mississippi landscape, a horizontal step on the model is a mile, and 5.4 minutes in its working is a day in the real world. The model brings into a one-third-square-mile outdoor laboratory the third largest river basin in the world, after those of the Amazon and the Congo. It is a basin that drains 41 percent of the continental United States and parts of two Canadian provinces—1,250,000 square miles, stretching from the Appalachians to the Rocky Mountains, Canada to the Gulf of Mexico.

German prisoners of war constructing the basin model under the supervision of engineers of the Waterworks Experiment Station [U.S. Army Corps of Engineers]

The model was constructed in the mid-1940s with the help of three thousand German prisoners of war for the U.S. Army Corps of Engineers through its principal research, testing, and development facility, the Waterways Experiment Station at Vicksburg. The Corps cleared and graded the model site, cutting and filling over a million cubic yards of earth to approximate the relief of the Mississippi basin, and installed eighty-five thousand feet of pipes to drain it. On this articulated surface the engineers positioned precast concrete segments of the model, one hundred feet square, on pilings that could control their elevation. The purpose of the model was to monitor the flow of water in the Mississippi River system, especially in times of flood, and to be a tool for testing and sanctioning future constructions. It was also intended to serve as a graphic and demonstrative display of river control for the public. Soil, translated into a coefficient of friction, is accounted for in textural additions to this vast concrete topography. And flowing water—monitored by specially designed, remotely controlled instruments and aided by pumps, valves, and pipes—represented the mighty Mississippi.

The basin model in operation during the 1973 flood [U.S. Army Corps of Engineers]

Today the model lies in a state of decay, apparently too costly to maintain. It served the Corps of Engineers until the 1973 flood of the Lower Mississippi, when it last assisted in determining which levees were in danger of being overtopped, how much they had to be raised, and how floodways had to be operated. The information furnished by the model over the period of its working life was found to be of "incalculable value" by the Corps. "The coordination of the vast and intricate plan for the whole basin would not otherwise be possible except by almost prohibitively laborious and time-consuming, as well as fallible, analytical methods."[1]

The Mississippi Basin Model, 1997

The overgrown topography of the basin model, 1997

The overgrown and crumbling model that the city of Jackson, its present owner, finds "degraded beyond repair," does not imply its rejection. It has merely moved from the weathering grounds on the eastern hills of the gulf coastal plain into the unfettered coordinates of cyberspace. The dematerialization of the Mississippi for the purposes of its control reaches back to well before the advent of computers. It goes back to early maps and surveys that sought to define its boundaries, to contain its horizon, first in the service of empire and then in the service of the empirical science of river hydraulics.

Henri de Tonty recounted the first claim on the Mississippi in his *Memoir* of 1693. He was a member of the earliest European expedition on record to travel the length of the Lower Mississippi, 140 years after De Soto reached it. Tonty was there on April 9, 1682, when upon planting a cross and raising the arms of France on the shores of the Gulf, "The Sieur de La Salle, in the name of His Majesty (Louis XIV), took possession of that river, of all rivers that enter it and of all the country watered by them." In the century that followed, explorers, naturalists, geographers, and army men sought to map the extent of the terrain that La Salle had so casually claimed.[2]

Guillaume de L'Isle's 1718 *Map of Louisiana* and John Senex's 1751 *Map of Louisiana and of the River Mississippi*, also showing "Habitations of the Indians, Promifeuous Nations, Nations destroyed," are among the most detailed and reprinted of early maps. Their project was not the Mississippi as such but the imagining of a continent. It was not long, however, before the river in itself began to draw the attention of surveyors. Maps drawn by French and British army officers showed "the general shape of the river and the location of permanent features such as tributaries, bluffs, and settlements" while also including the "character of land and vegetation adjacent to the river." The Ross Map (1765), the Pitman Map (1768), and the Collot Map (1796) are some of the more popular of these "reconnaissance" maps of the Lower Mississippi. They used compass bearings and estimated distances, far from the methods of the exact science that was to follow. Somewhat different in measure and features but dating from the same period is the Wilton Map (1774). It is based on a cadastral survey, its primary purpose being the portrayal of property lines. Despite their inaccuracies by later standards, the tracings of the river in these maps, particularly the Ross Map, are the earliest to feature in overlays done in this century to track and understand the progressive channel changes of a shifting Mississippi.[3]

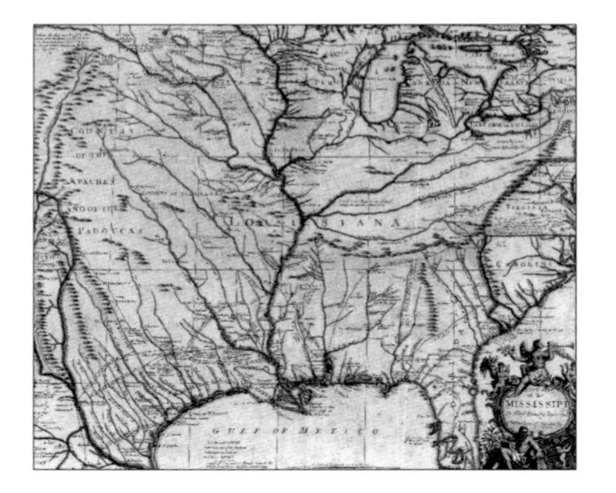

John Senex, *A Map of Louisiana and of the River Mississippi*, 1751 [The Historic New Orleans Collection, accession no. 1972.8]

Lieutenant Ross, *Course of the River Mississippi from the Balise to Fort Chartres*, 1765 [Annenberg Rare Book and Manuscript Library, University of Pennsylvania]

General shape and features were adequate for colonization, but sorely inadequate for the navigation of a river that kept shifting even as it was fast becoming a major outlet for the vast commercial empire to which La Salle laid claim. Its status as an outlet was plotted in 1699 when Pierre Le Moyne d'Iberville, founder of the first French settlement in Louisiana, entered the Mississippi from the Gulf and was given a letter by Indians written by a member of La Salle's party. The letter confirmed that the explorers were on the same river, completing a circuit that connected the vast resources of the northern territories traveled by La Salle with the markets brought by Iberville from the Gulf. It was not until 1820, however, eight years after the Lower Mississippi became wholly United States territory when Louisiana was admitted as the eighteenth state, that the first complete survey of the Lower Mississippi for the purposes of navigation was commissioned by the U.S. Congress. It marked the beginning of federal involvement in this landscape, with the work being carried out by the recently merged Corps of Topographical Engineers and the Army Corps of Engineers. But it was the survey commissioned by Congress in 1850 with the added purpose of flood control that is a landmark for going beyond describing shape and features.

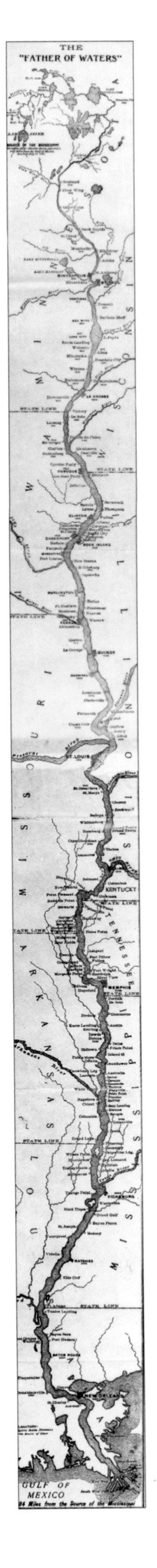

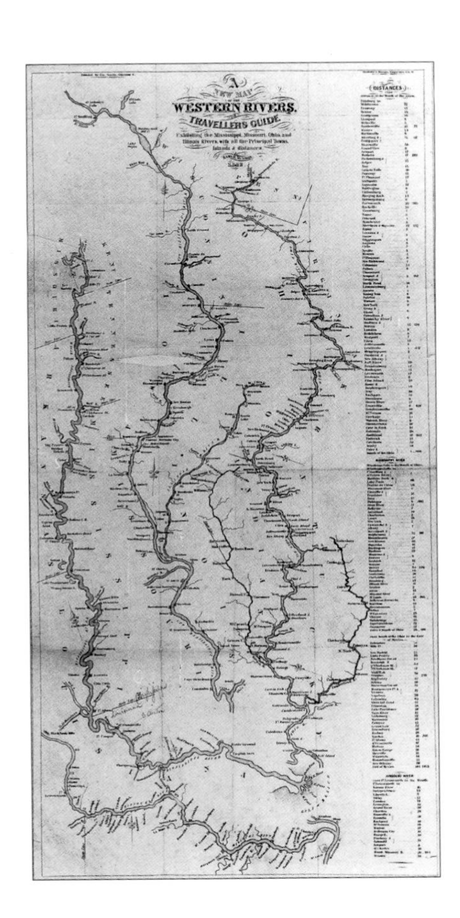

Marie Adrien Persac, *From Natchez to New Orleans: Norman's Chart of the Lower Mississippi*, 1858 [The Historic New Orleans Collection, accession no. 1947.1]

Map of the Father of Waters, from Captain Willard Glazier, *Down the Great River*, 1888

S. B. Munson, *A New Map of Western Rivers, or Travellers Guide*, 1848 [The Historic New Orleans Collection, accession no. 1981.266]

The Delta Survey, as it is called, included a "Map of the Alluvial Region of the Mississippi" and a massive report on the "Physics and Hydraulics" of the river published in 1861. It was the first scientific study of the Mississippi River and was to influence policies and constructions for decades, until events led to the questioning of many of its assumptions. The Delta Survey marked a shift. Like its predecessors, it was not merely descriptive. It was also projective, its information being a purposeful abstraction of the world it portrayed. But it goes further than its predecessors. It places this information at the service of river hydraulics, an empirical science that views the Mississippi as a subject of controlled experimentation. This science seeks information that is not merely directed to understanding river dynamics but also to carrying out engineering operations that control the river for study, operations such as levee building, dredging, reservoir and floodway construction, stabilization of banks. The purpose of this science is "to determine the laws governing the flow of water in natural channels and to express these laws in new formulae which could safely and readily be used in practical application." [4]

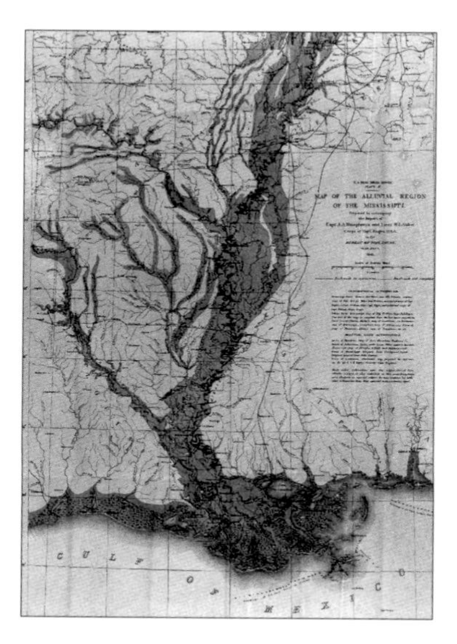

"Map of the Alluvial Region of the Mississippi," from the U.S. Mississippi Delta Survey report by A. A. Humphreys and H. L. Abbot, 1861 [The Historical Society of Pennsylvania]

In 1879 Congress authorized this experimentation with the creation of the Mississippi River Commission. It consisted of seven members appointed by the president of the United States—three from the Army Corps of Engineers, one from the coast and geodetic survey, and three from civilian life, two of whom are civil engineers. "The Commission was charged with the preparation of surveys, examinations, and the preparation and consideration of plans to improve the river channel, protect and stabilize the river banks, improve navigation, prevent destructive floods, and promote and facilitate commerce and the Postal Service," according to a history of the commission. It ushered in a new era of scientific optimism. [5]

Early levee construction

A giant willow mat, a predecessor of today's concrete revetments, being laid in place to prevent bank erosion [U.S. Army Corps of Engineers]

For forty-eight years, until the 1927 flood, surveys carried out by the commission were directed toward testing a "levees-only" idea. This idea was backed by a popularly accepted theory of river hydraulics: that if the river was prevented from spreading either through distributaries or through wider channels, then the resulting increase in volume and velocity in a channel would have two effects. First, it would scour the river bed, thereby deepening the channel for navigation; and, second, it would prevent sediment from settling and choking the channel. This was a "self-dredging" theory of alluvial

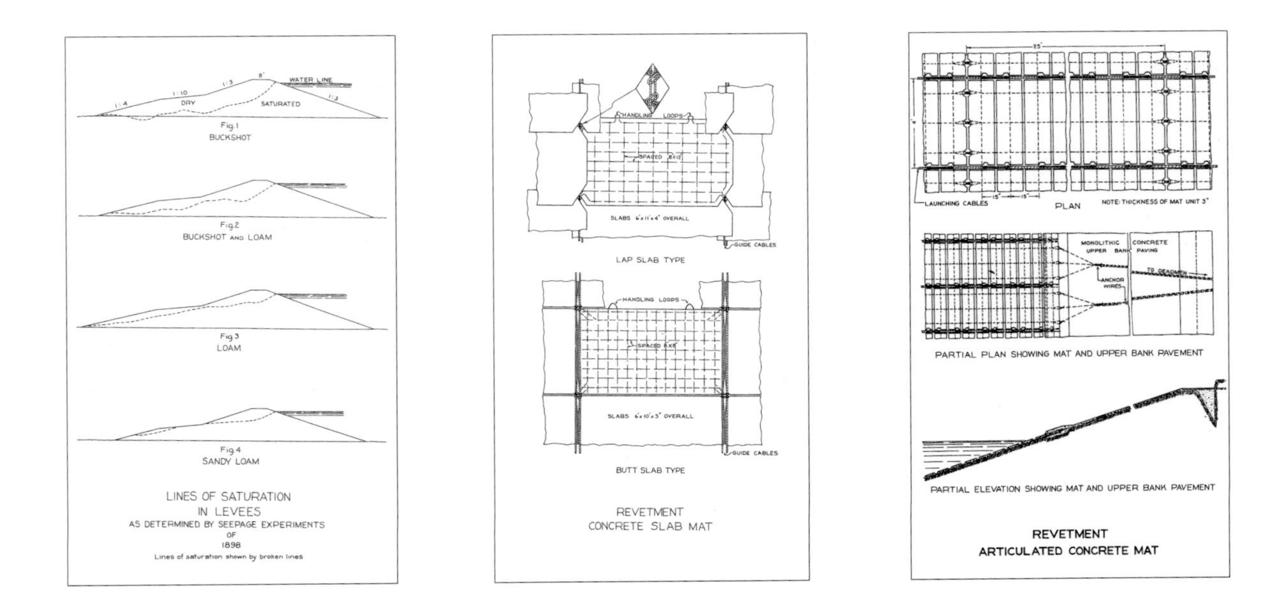

Study drawings of levee and revetment construction, from D. O. Elliott, *The Improvement of the Lower Mississippi River for Flood Control and Navigation*, 1932

rivers developed by the seventeenth-century Italian engineer Domenico Guglielmini from observations of the Po River.[6] Numerous rivers around the world, notably the Yellow, the Tigris and Euphrates, and the Nile, have been confined by levees (also called dikes) for thousands of years. But these were built to protect the land rather than to speed a river. Levee building, prior to the Delta Survey, had apparently little to do with the empirical science of river hydraulics.

The illustrious engineer James Eads used Guglielmini's theory to advantage when he designed his famous jetties—walls of stone and willow perpendicular to the flow to narrow a channel—in order to open the passes of the Lower Mississippi Delta. The three major passes of the delta—South Pass, Southwest Pass, and Pass à Loutre (also known as North Pass)—kept silting with the spread and loss of velocity as the Big Muddy entered the Gulf. Eads made a historic contribution to navigation. The theory also tended to support cutoffs—channels that cut across the necks of an alluvial river's sweeping horseshoe curves. Straightening the river with cutoffs increased its slope and hastened its waters and sediment to the Gulf. But levees-only was the major and most controversial practical application of Guglielmini's theory. Confinement by levees, it was believed, "would cause the river to scour out the channel enough to accommodate floods." Almost by default Guglielmini's theory was opposed to a theory of outlets that backed the relatively simple belief that removing water by either reservoirs or floodways would lower flood levels. Surveys therefore also sought to present information on outlets and to describe their nature either in support of a levees-only theory or to argue against the dominance of levees. But the case for outlets could only be speculative while the levees-only policy held demonstrative sway, pushing for information on river discharges, sandwave movement, cross-section soundings, ratio of sediment to water, and so on, but also for soil subsidence under levees, seepages, costs of levee construction, and so on.[7]

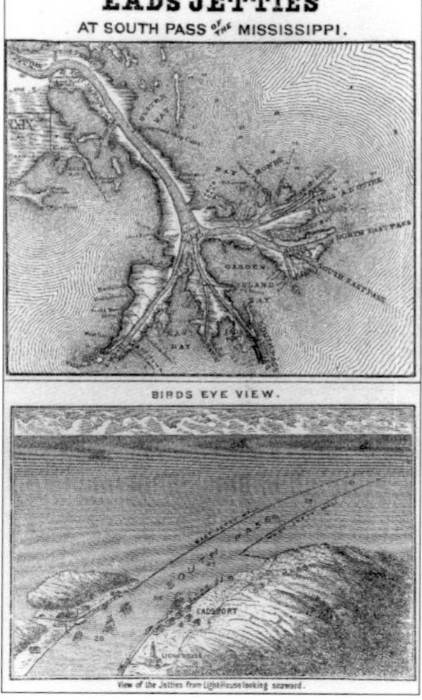

J. O. Davidson, *Eads Jetties at South Pass of the Mississippi*, 1883
[The Historic New Orleans Collection, accession no. 1950.58.3]

This was an era that could afford to straighten the river in maps. The Lower Mississippi, with its initial north–south run leading into an east–west stretch before flowing into the Gulf at an indecisive orientation, challenged efficient framing in navigation books and government documents. Strategic straightening of the river channel was experimented with in maps in order to accommodate the detailed channel efficiently on one sheet, or several consecutive pages of a report. It was a device common in early cultural maps of the Mississippi, such as *A New Map of Western Rivers, or Travellers Guide* by S. B. Munson in 1848, *From Natchez to New Orleans: Norman's Chart of the Lower Mississippi* by Marie Adrien Persac in 1858, and "Father of Waters" by Captain Willard Glazier in 1888. The realignment of the Mississippi in these and later maps by the Corps, if not encouraged, accommodated the levees-only policy.[8]

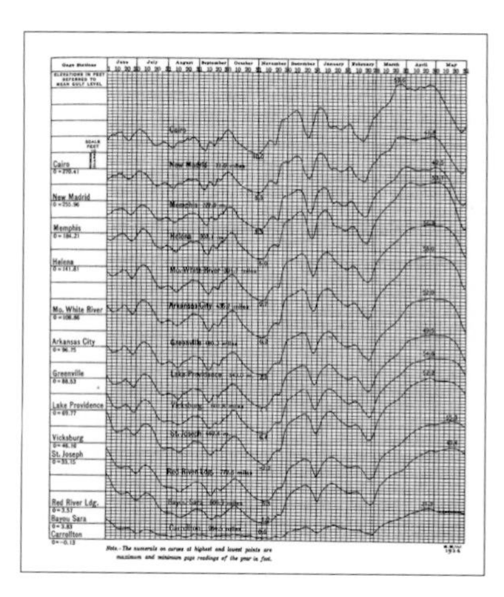

A hydrograph recording (1921–22) showing the considerable fluctuations in water levels of the Lower Mississippi River
[U.S. Army Corps of Engineers]

Then came the 1927 flood, "the most disastrous in the history of the river." It inundated twenty-three thousand square miles of the valley that had been protected from the Mississippi by more than fifteen hundred miles of levee. Of the thirteen crevasses—breaches in the levees that gouged land to depths of as much as one hundred feet, increasing the destructive force of the torrent—in that flood, one was caused by dynamite. It was found necessary to blast the levee south of New Orleans to protect the city from inundation. It made the case for outlets. As historian Albert Cowdrey put it, "A policy had been breached, and the pouring waters were sweeping an era away."[9]

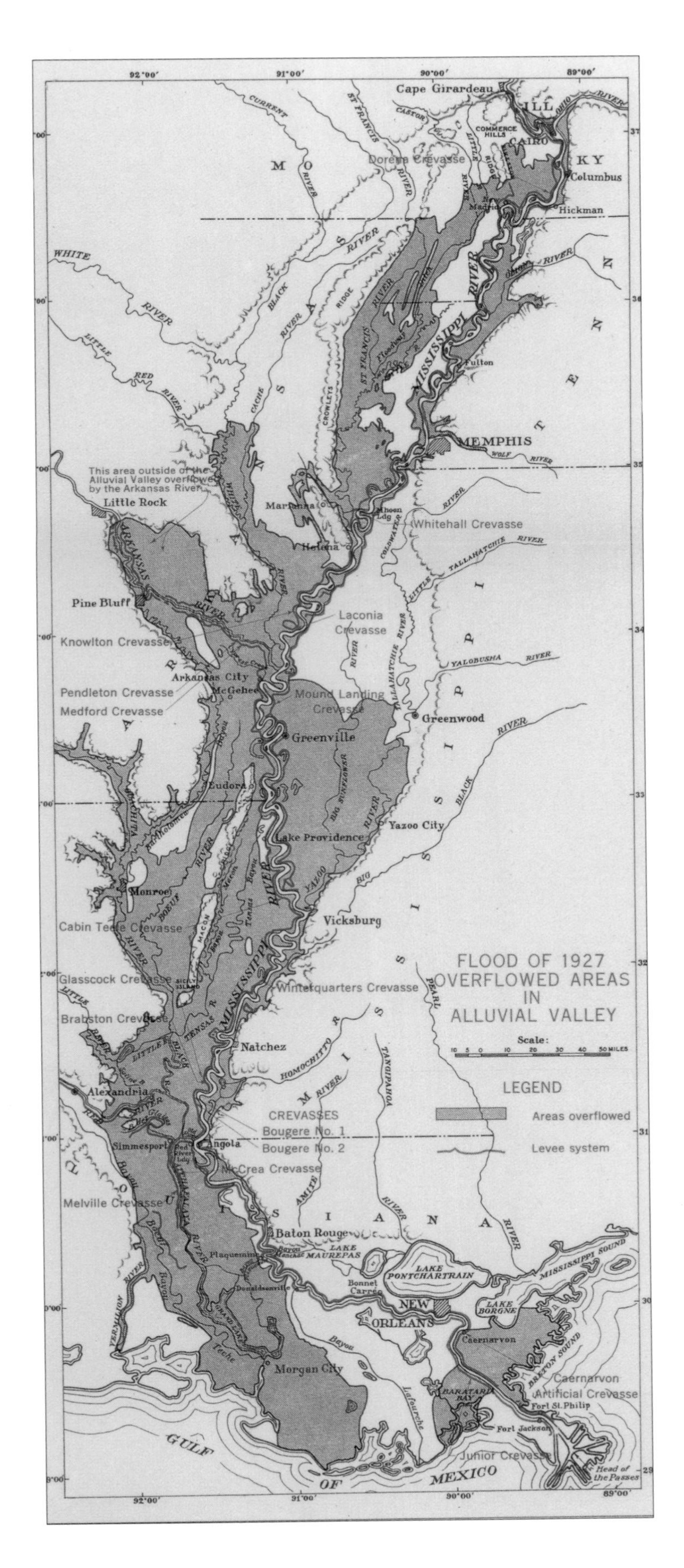

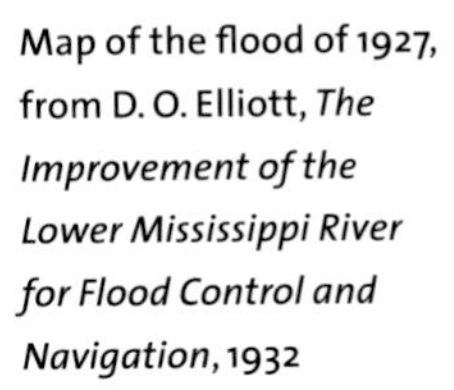
Map of the flood of 1927, from D. O. Elliott, *The Improvement of the Lower Mississippi River for Flood Control and Navigation*, 1932

The view that gained favor was that "the river needs more room which should be given to it laterally rather than vertically." It met with immediate approval from people who "did not know what they wanted but most of them were convinced that they wanted something different from what they had been getting." Their patience had been tested by 17 floods and more than 730 crevasses totaling 125 miles in the 48 years since the Mississippi River Commission began the levees-only experiment.[10]

With outlets began a new experiment. The "old erroneous 'levees only' policy of flood control was abandoned without a blush or seeming regret for the reckless loss of life and property and unprecedented suffering which these same authorities had foisted upon citizens who believed in them for more than a hundred years." Project Flood, as the new experiment was called, involved considerable investment in the construction of safety valves, controlled spillways, fuse-plug levees, and eventually the Mississippi Basin Model. The empirical science of river hydraulics had a new playing field.[11]

Left:
Map of the progressive channel changes of the Lower Mississippi River, from D. O. Elliott, *The Improvement of the Lower Mississippi River for Flood Control and Navigation*, 1932

Center:
Map showing trans-valley cross sections, from Harold N. Fisk, *Geological Investigation of the Alluvial Valley of the Lower Mississippi River*, 1944

Right:
"Distribution of Project Flood," from P. A. Feringa and W. Schweizer, *One Hundred Years Improvement on the Lower Mississippi River*, 1952

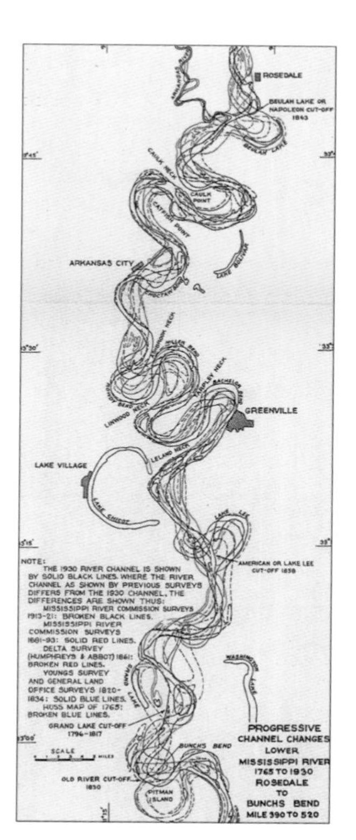

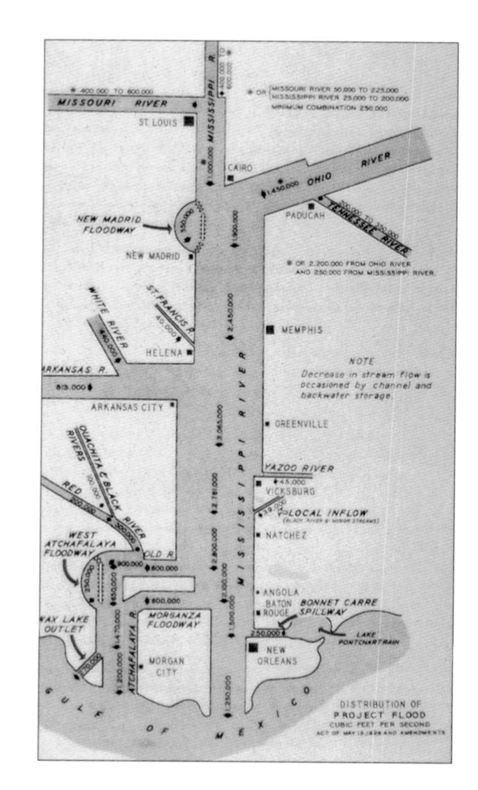

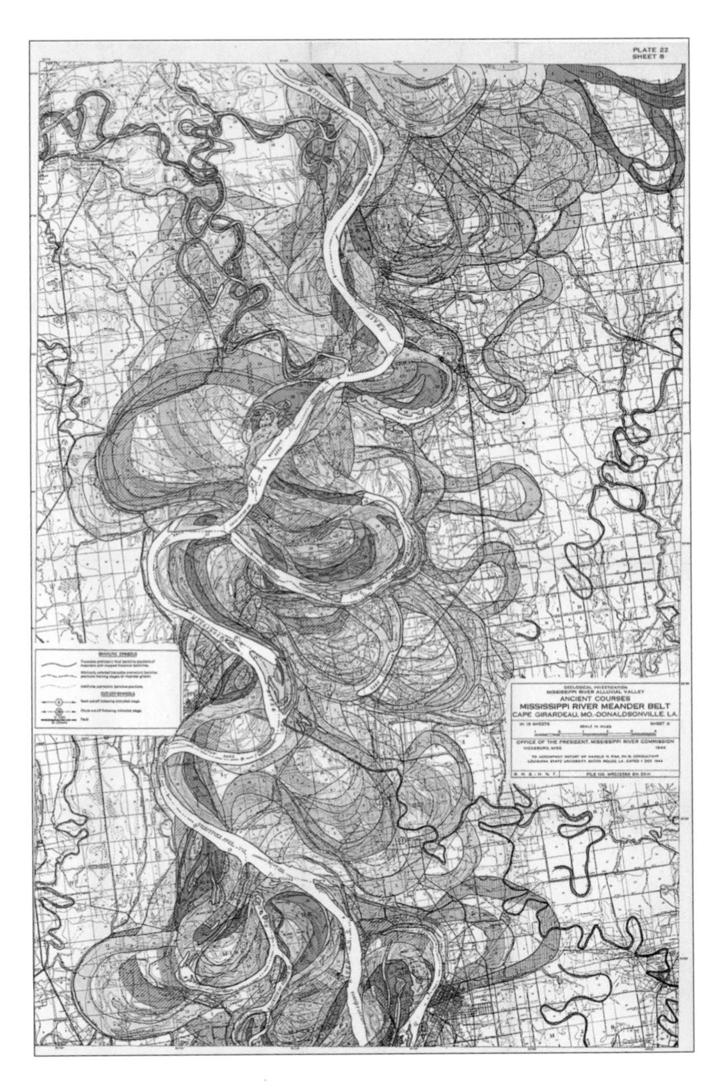

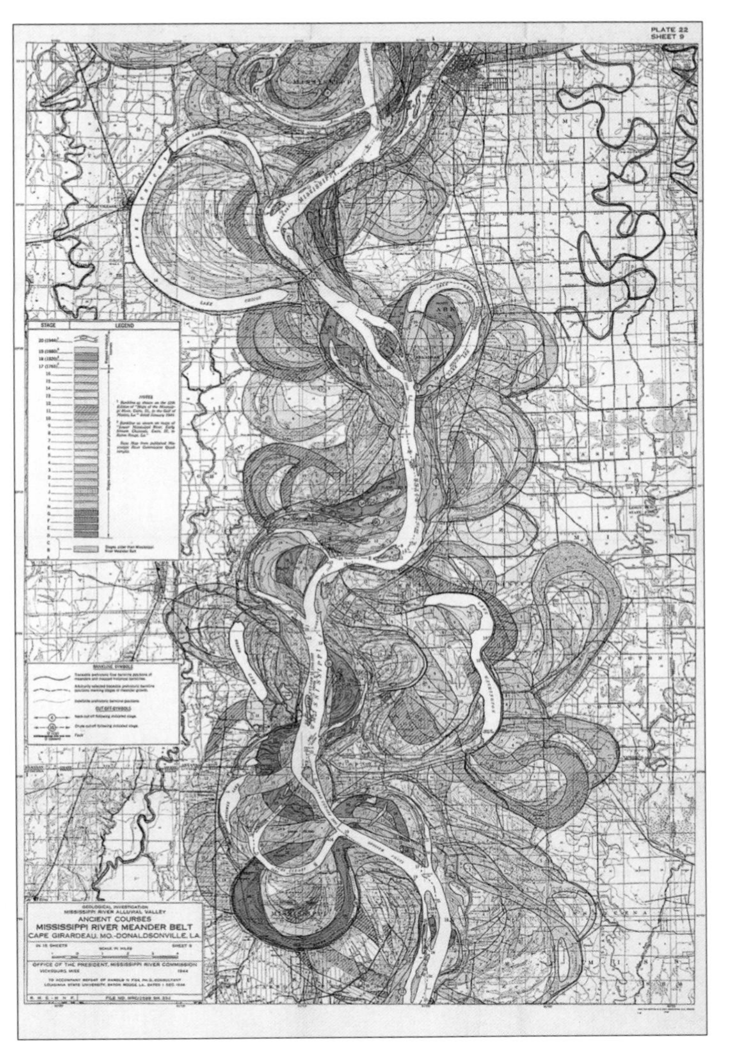

Map of the ancient courses of the Mississippi River meander belt (two parts), from Harold N. Fisk, *Geological Investigation of the Alluvial Valley of the Lower Mississippi River*, 1944

Whereas each flood was the testing ground for the levees-only approach, the Mississippi Basin Model served as a testing ground for Project Flood. Indeed, the whole concept of a Waterways Experiment Station Laboratory was conceived to relieve the actual valley of some of the ill-fated consequences of experiments by the Corps. More important, however, the Mississippi Basin Model provided the new data introduced by Project Flood. Designs were to be measured against the maximum flood predicted possible—a hypothetical flood rather than the next flood. The Mississippi Basin Model could run the worst-case scenario, testing outlets even before they were built.

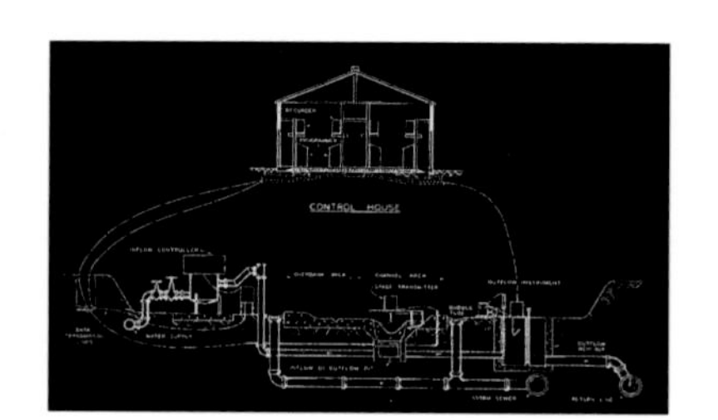

Water supply and drainage diagram of the Mississippi Basin Model [U.S. Army Corps of Engineers]

Detail of working drawing of the drainage system of the Mississippi Basin Model [U.S. Army Corps of Engineers]

On the two hundred acres at Clinton is the empire that La Salle hardly knew when he "took possession of that river, of all rivers that enter it and of all the country watered by them" in 1682. It features flood—"great floods of the past and the possibly greater floods of the future can be created . . . in accurate miniature." Flood is perhaps the least of the reasons why La Salle desired possession. But floods come with the Mississippi basin and not merely as the inundation of inhabited land. In Sites 1, 2, 3, and 4, as in Site 0, floods construct the Mississippi both in the imagination and on the ground.[12]

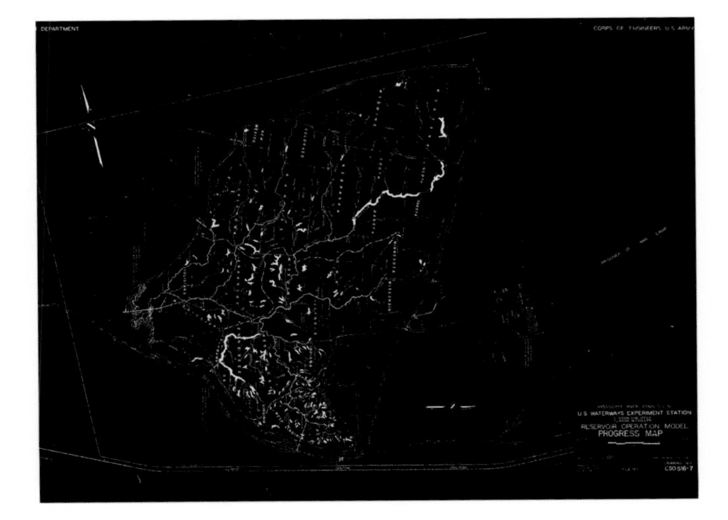

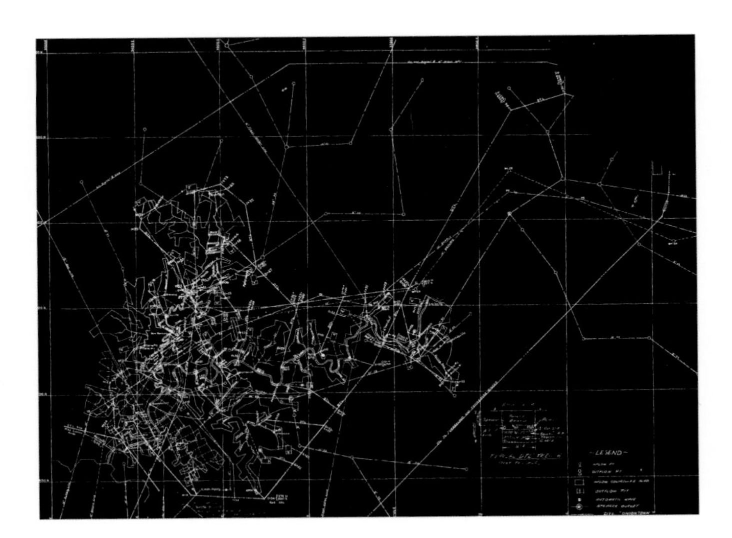

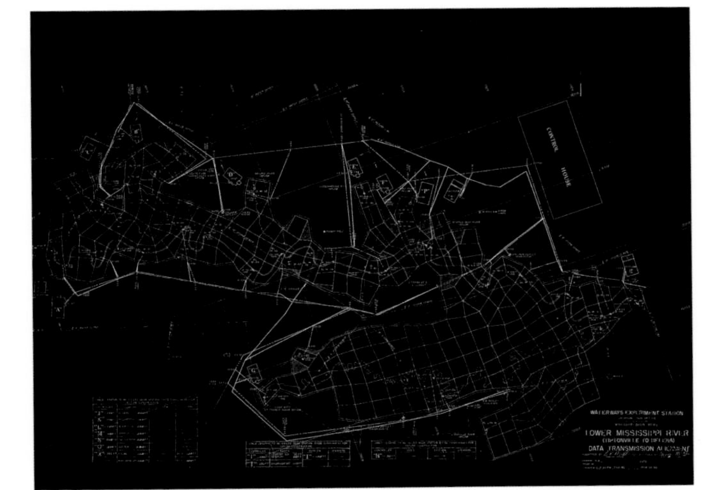

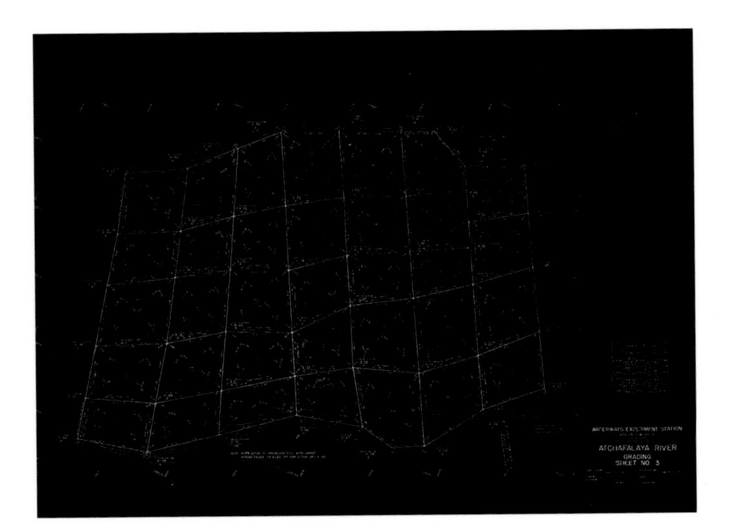

Layout drawings of the Mississippi Basin Model

[U.S. Army Corps of Engineers]

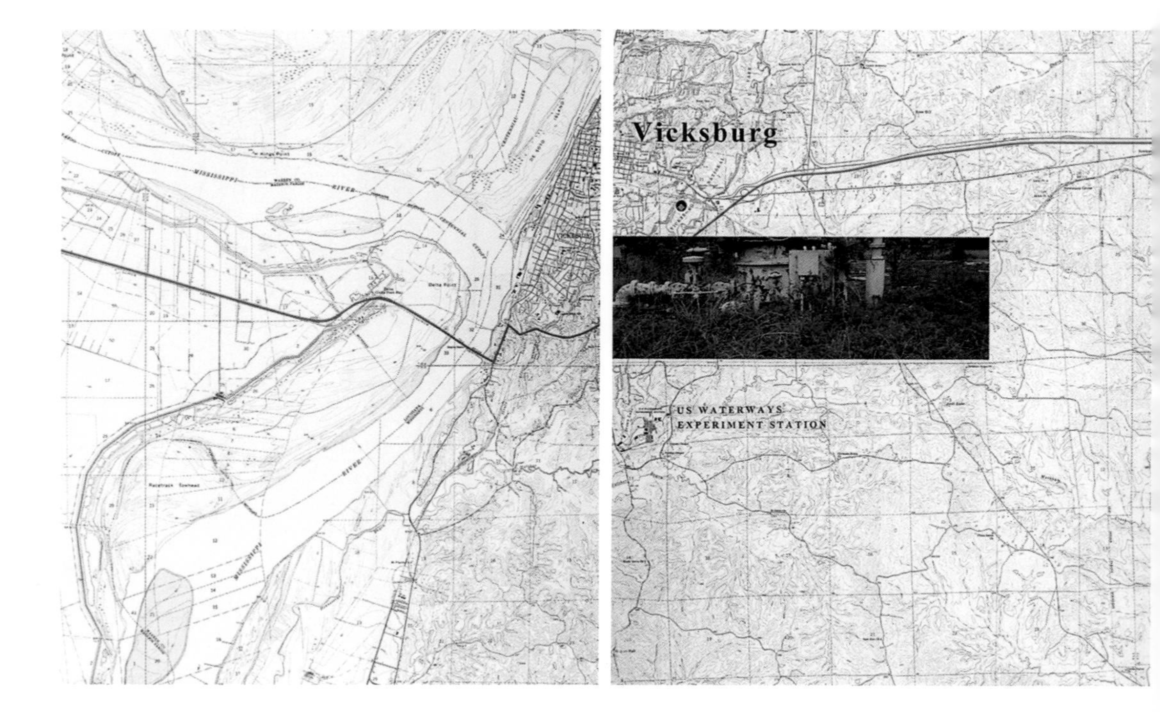

On a Concrete Surface
(color photo prints on USGS maps, each panel 15" x 17")

Fenced with wire and surrounding a watchtower on the brink of collapse is a vast and weathered concrete field. Much of it is covered with vegetation and an accordion-like wire mesh about five inches high resembling vegetation. Running across the surface in river-like fashion are deep channels edged by steep embankments. These channels vary in depth from an inch to a couple of feet. Crisscrossing this sinuous landscape orthogonally are the white chalky joints between the individual units of the molded concrete, rusting pipes that run a few inches above the surface, and drains that end in open manholes. Occasionally, where the joints have eroded, one gets a glimpse of the ground upon which the concrete units rest on supports. It is empowering to know that a step in this ruinous landscape takes you a mile across the Mississippi Basin; the channels, where they are a foot deep, take you one hundred feet down
into the Mississippi.

SITE 1

MEANDERS

GREENVILLE BENDS / YAZOO DELTA

Delta Crossings
(acrylic and pastels on paper,
each panel 12" x 9")

BLUES MEANDERS

Past Cape Girardeau, Missouri, moving sluggishly across the nearly flat surface of a huge trough of alluvium 30 to 90 miles wide and 600 miles long, the meandering Mississippi carries a self-organizing power, much of which lies in the bend. Bends, shaped more like horseshoes than hairpins, force the sedimentation of one bank (a point bar) and the erosion of the other (a cut bank). This continuous movement of sediment thereby extends the sweep of meanders until two adjoining bends meet when the neck that separates them gets too narrow to hold and is eventually cut off, and the water jumps over it. But what is lost in length with a cutoff is regained elsewhere as the alluvium surface of the valley, sloping a mere 290 feet in 600 miles—a slope of 1 in 11,000—does not easily accommodate changes in velocity. Instead the momentary increase merely pushes the formation of another bend elsewhere.

The Greenville Bends [U.S. Army Corps of Engineers]

For big boats the doubling of the distance to the Gulf caused by bends is inefficient, especially when, as described by Mark Twain, the bends can be so deep that "if you were to get ashore at one extremity of the horseshoe and walk across the neck, half or three-quarters of a mile, you could sit down and rest a couple of hours while your steamer was coming around the long elbow at a speed of ten miles an hour to take you on board again." To the man in a canoe who floats from Lazy Man's Landing downriver thirty-five miles along the famous Greenville Bends to Greenville, the bends are an assistance, for to return, he does not paddle back up. Instead he floats downstream, pulls his canoe across the narrow neck, floats downstream again, repeating the act until he gets back home, upriver.[1]

Greenville is the most prominent town of the Yazoo Delta. Simply referred to as the delta, this leaf-shaped region, 220 miles long and 70 miles at its widest, is the basin of the Yazoo River. It is framed by loess bluffs on the east and the earthen embankments that run along the Mississippi River on the west. To many this area is the "South's South," "Mississippi's Mississippi," famous for its politics of race and poverty.[2] But it is as famous for its rich soil, ideal for growing cotton, and the Delta blues, a music with its roots in the chants of slaves, their release from bondage when they cleared and drained this backswamp of the Big Muddy and then worked the cotton plantations before the Civil War.

What the bends do for a meandering Mississippi the blues do for the Delta blues: they make this music—one of America's originals and the forerunner of jazz—hard to capture in notes and bars. One commentator describes the blues as "at once a way of life, a variety of music, a poetic movement, a state of mind, a folkloric tradition, a moral attitude, and even a kind of spontaneous intuitive critical method. Most commentators agree that it somehow repels all efforts to harness it too tightly in any definition. However, this very indefinability, this many-sided elusiveness, is itself revealing about its fundamental character." [3]

Levee construction, 1800s [U.S. Army Corps of Engineers]

While the blues give to the Delta blues the twists, turns and quavers, swooping dips and sudden climbs, the attack and release of notes, and spontaneous embellishments that make the music so elusive, they also give it an emergent, ordinary everydayness. There are high-water blues, boll-weevil blues, cottonfield blues, dirt-road blues, penal-farm blues, turtle-dove blues, and, today, despite pressure from tourists to the Yazoo Delta who want to hear the blues of a past era of cottonpickin', there are drug blues, talking-back blues, cyberspace blues. In the Yazoo Delta, Charlie Patton's "High Water Everywhere" ("Lord the whole roun' country Lord, river is overflowed / Lord the whole roun' country, man it's overflowed / I would go to the hilly country, but they got me barred") is not merely about the loss of life and property in the disastrous 1927 flood, nor is it merely about the limits imposed upon him as a black in the South: he is giving voice to a land constructed by a meandering Mississippi, bonded but not bound. [4]

Dockery Farm, workplace of Charlie Patton and a number of other blues musicians

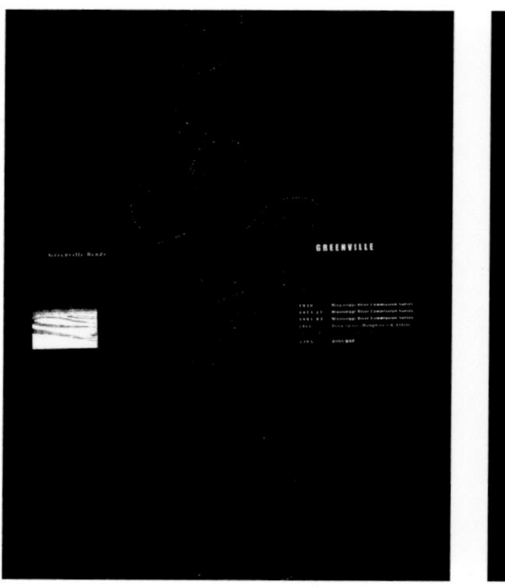

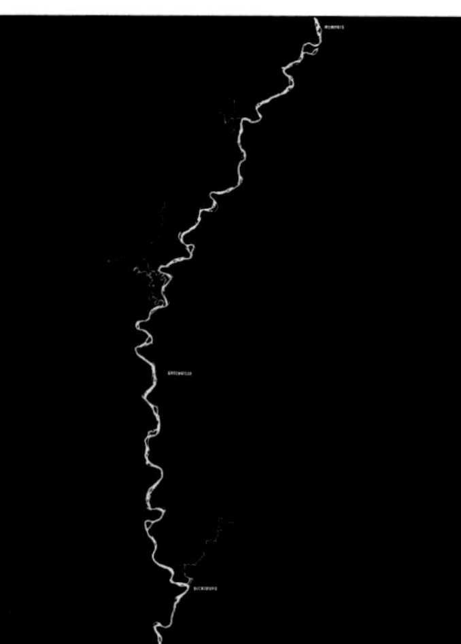

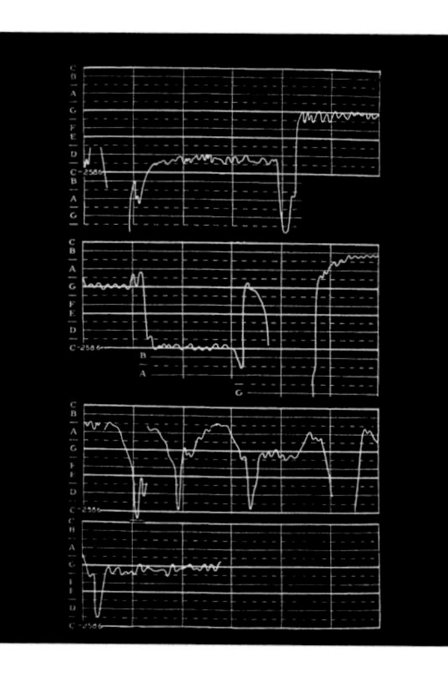

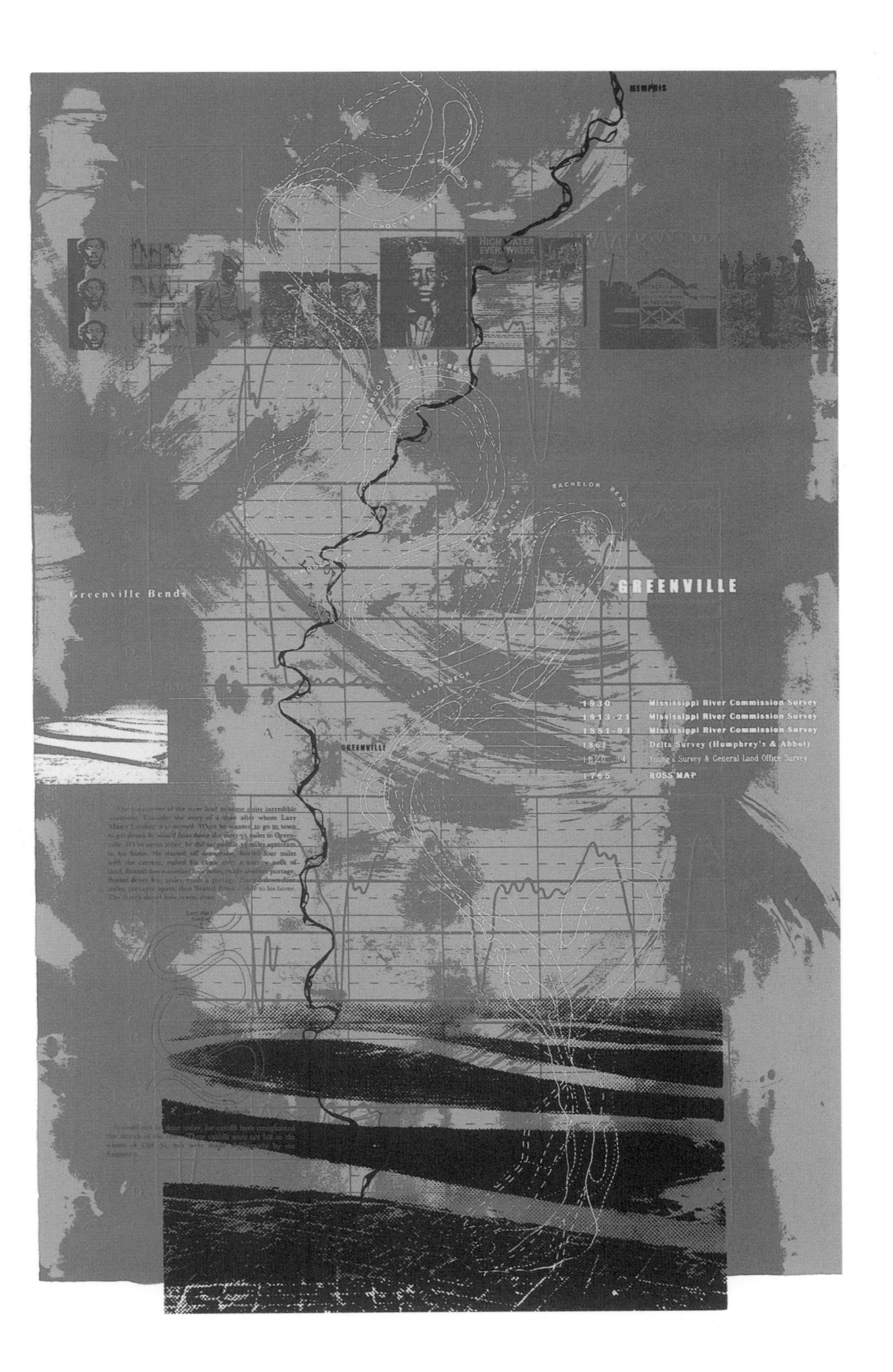

Blues Meanders (screen print on paper, 30" x 44")

Greenville Bends

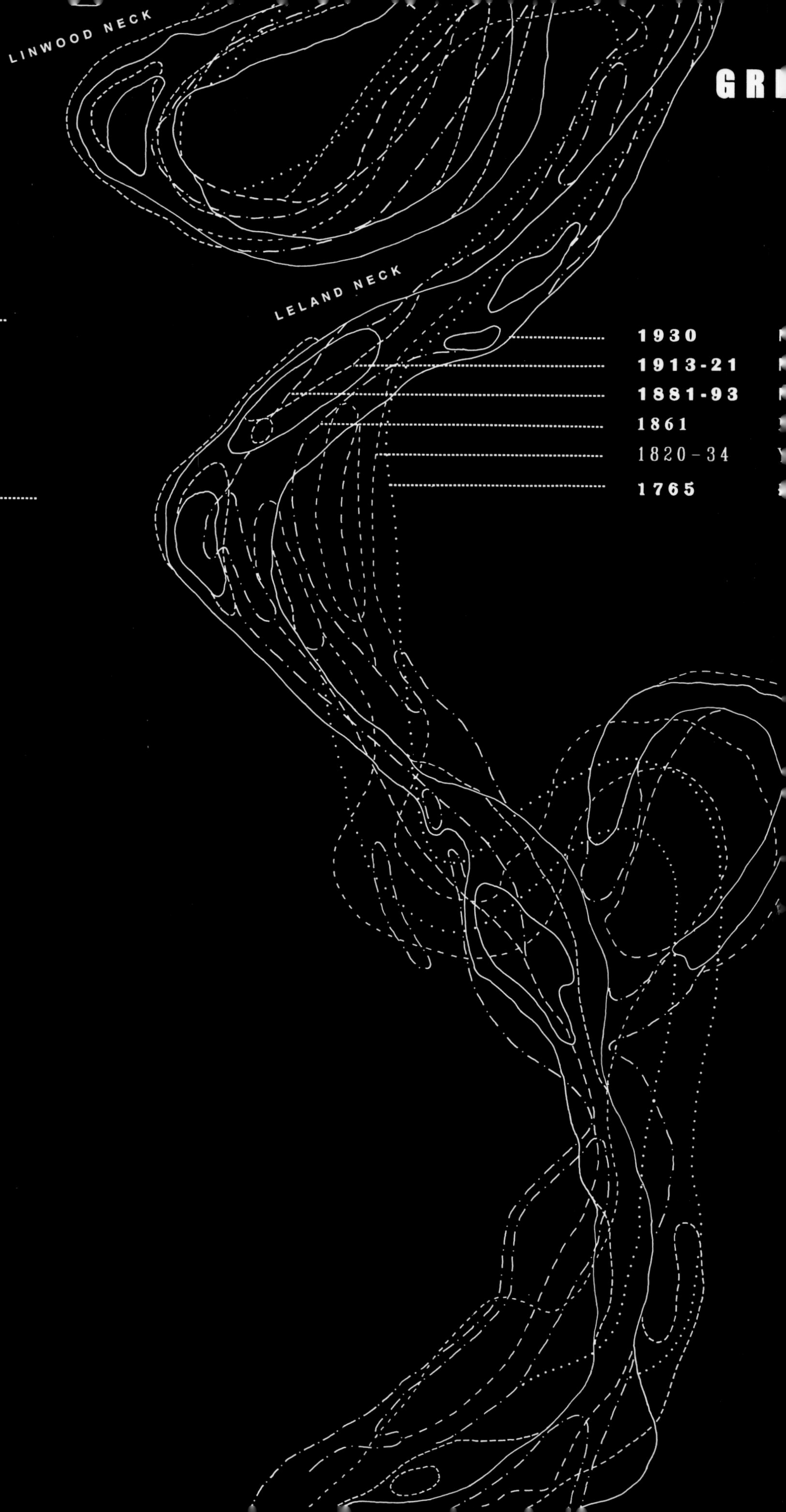

LINWOOD NECK

LELAND NECK

1930

1913-21

1881-93

1861

1820–34

1765

GR

LINWOOD NECK

LELAND NECK

GREENVILLE

1930

1913-21

1881-93

1861

1820 34

1765

The contortions of the river lead to some quite incredible situations. Consider the story of a man after whom Lazy Man's Landing was named. When he wanted to go to town to get drunk he would float down the river 35 miles to Greenville. When again sober, he did *not* paddle 35 miles upstream to his home. He started off *downstream,* floated four miles with the current, pulled his canoe over a narrow neck of land, floated down another four miles, made another portage, floated down four miles, made a portage, floated down four miles, portaged again, then floated down a mile to his home. The sketch shows how it was done.

Lazy Man's Landing

Greenville

It could not be done today, for cutoffs have straightened this stretch of the river. These cutoffs were not left to the whims of Old Al, but were made deliberately by the Engineers.

ENGINEERED CURVES

It is conceivable that if the Mississippi is restrained, for example, by embankments, then an induced and controlled cutoff will shorten it, and consequently increase the speed of its flow. The embankments will prevent the river from compensating for the loss of length with another bend or a flood. Such a cutoff can eliminate as much as thirty miles of the Mississippi. For tows, the giant barge assemblies the size of six football fields laden with seventy-two thousand tons of goods, that means two to three hours saved. For those concerned with preventing floods it means speeding up sediment that would otherwise settle and raise flood levels. For those with land along the potential cutoff it means soaring land values, although their good fortune is countered by the ill fortune of those who will be deprived of river frontage. But it is a matter of power, or, as Mark Twain suggests, deviousness easily executed by cutting a "little gutter" to allow the River King to do the rest. Add to these benefits the widely held belief that rivers mature as they shorten their route to the sea and that the Mississippi in this regard is only an infant river at the very early stages of its development, and one has a very powerful case for engineering cutoffs.

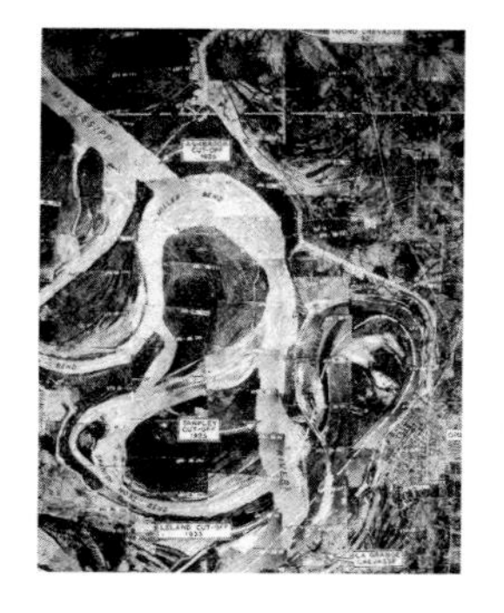

Photomosaic of the Greenville Bends soon after the cutoff, from Harold N. Fisk, *Geological Investigation of the Alluvial Valley of the Lower Mississippi River*, 1944

According to Twain, in the 176 years before he wrote *Life on the Mississippi*, the Lower Mississippi was shortened (and shortened itself) by 242 miles. His projection in 1874 on this basis makes the point of cutoffs: "seven hundred and forty two years from now the Lower Mississippi will be only a mile and three quarters long, and Cairo and New Orleans will have joined their streets together." Many would like to see the Lower Mississippi disappear, if not materially, at least from worry. In the 1930s and '40s, following the disastrous 1927 flood that inundated Greenville, cutoffs were authorized by the Mississippi River Commission with the intention of eliminating the erosion of embankments, speeding the transportation of sediment, and dropping flood heights. Their efforts shortened the river by 152 miles.[5]

Greenville cutoff, 1930s
[U.S. Army Corps of Engineers]

Cutoffs in this period also eliminated the Greenville Bends and with them the self-organizing power of the meander that made the Yazoo Delta, the small boats of small enterprises that cannot ply the swifter river, and the fish that cannot find a "quiet spot to breed" in the more turbulent waters.[6] They have been replaced by the organized curves of the Army Corps of Engineers, curves held in place by concrete paving. The

Corps' purpose is to hasten water, silt, sewage, and vessels to the Gulf. Greenville today sits on an abandoned arm of its once famous bends, which have been reduced from a sweeping fifty-one miles to a relatively straight but distant stretch of twenty-four miles. In flood, however, or at least in the constant threat of it, one hears the voice of the bend, which like the blues expresses resistance to bondage. Despite the tight controls of embankments, the river since the last cutoff has regained more than fifty miles.

The intersection of Lake Ferguson (right) and the Mississippi River in its new location following the Greenville cutoff

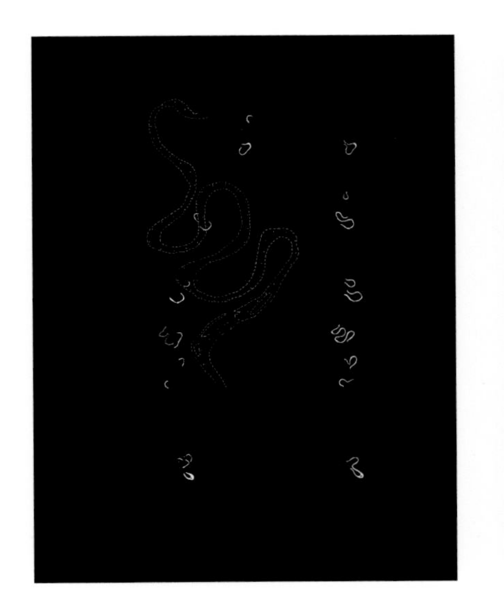

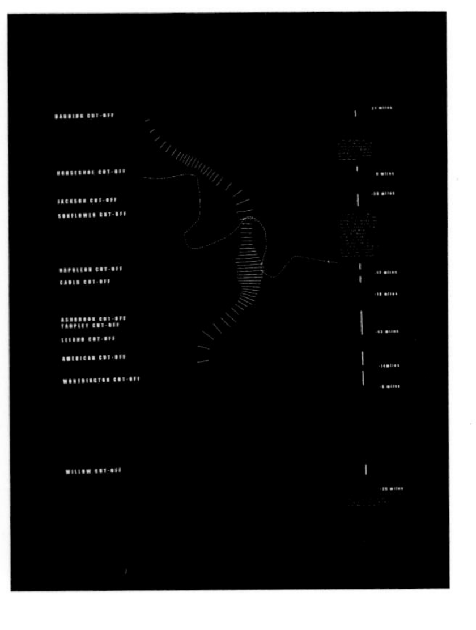

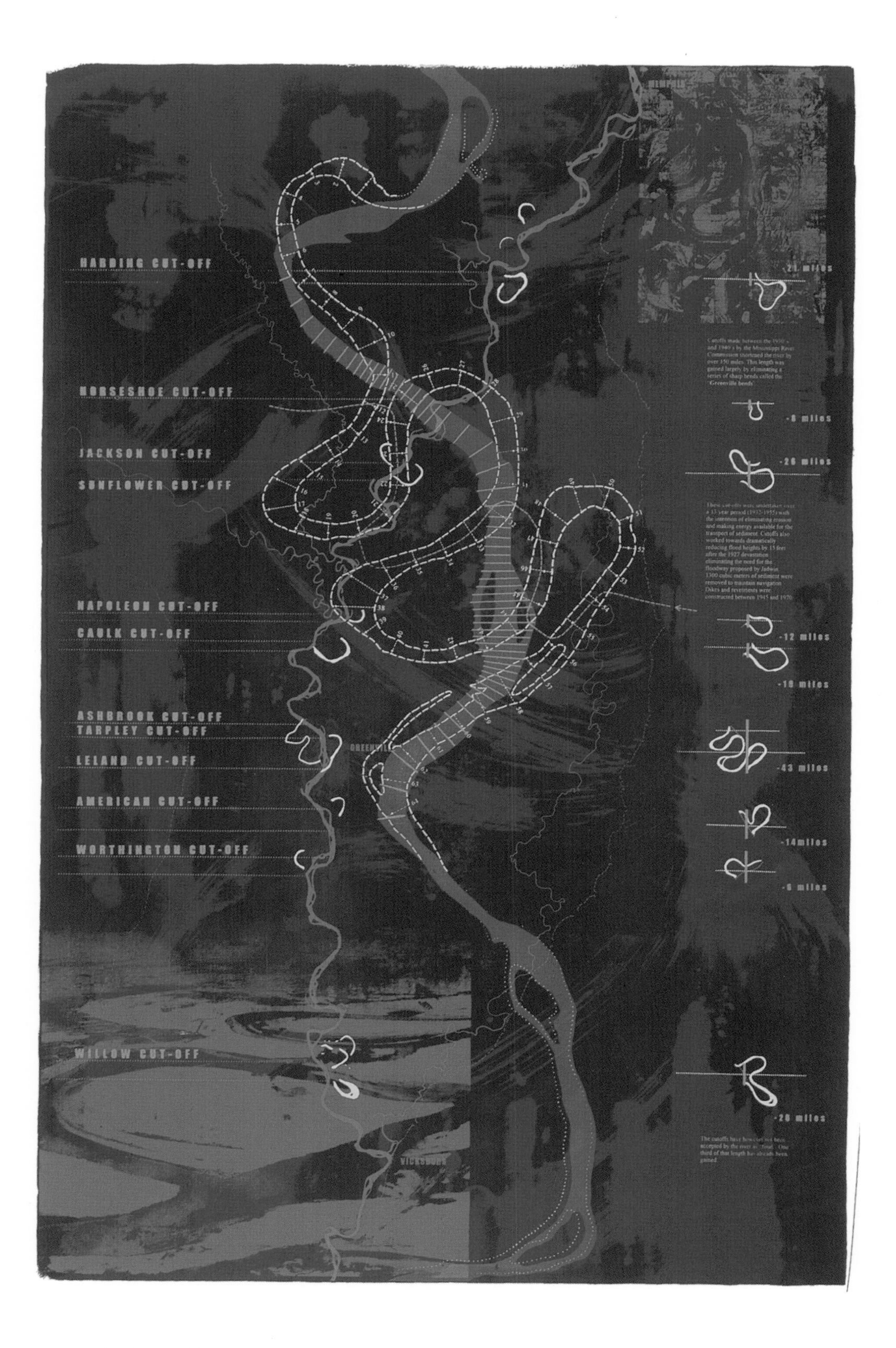

Engineered Curves (screen print on paper, 30" x 44")

Cutoffs made between the 1930's and 1940's by the Mississippi River Commission shortened the river by over 150 miles. This length was gained largely by eliminating a series of sharp bends called the "**Greenville bends**"

These cut-offs were undertaken ove a 13 year period (1932-1955) with the intention of eliminating erosion and making energy available for the transport of sediment. Cutoffs also worked towards dramatically reducing flood heights by 15 feet after the 1927 devastation eliminating the need for the floodway proposed by Jadwin. 1300 cubic meters of sediment were removed to maintain navigation. Dikes and revetments were constructed between 1945 and 1970

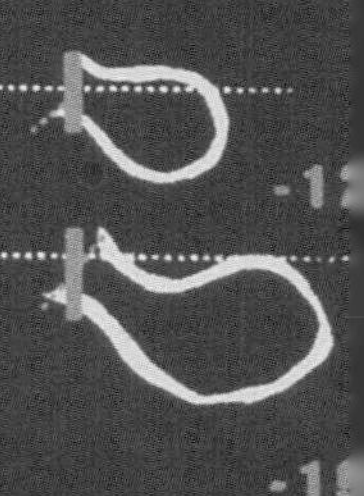

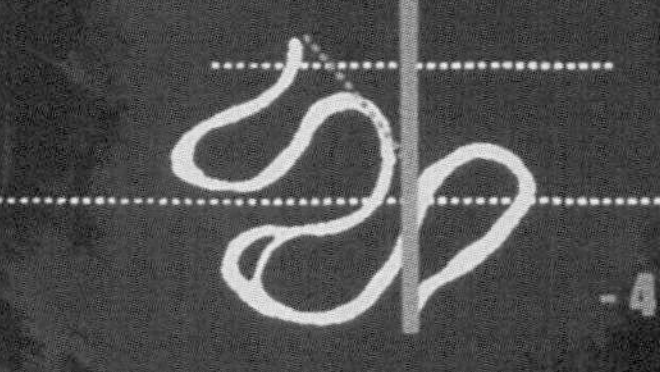

DEPOSITING DEPTH

The meandering Lower Mississippi, which moves more with the kinetic energy of its waters than the potential energy of its slope to the Gulf, spreads itself easily, particularly with a surge from a snowmelt or a storm. Although its spread is liable to be destructive, it also constructs by depositing the loads of sediment that it has eroded from the vastness of its basin, banks, and beds. Deposits diminish with distance from the stream, creating embankments that slope gradually into backswamps of finer silt and clay. Over a period, with all the shifts in the positions of bends, an alluvial ridge called a meander belt can form within the alluvial valley. It can be as much as fifteen feet high and fifteen miles wide. But a strong surge may change the course of the stream altogether, to abandon a meander belt and begin the formation of a new one.

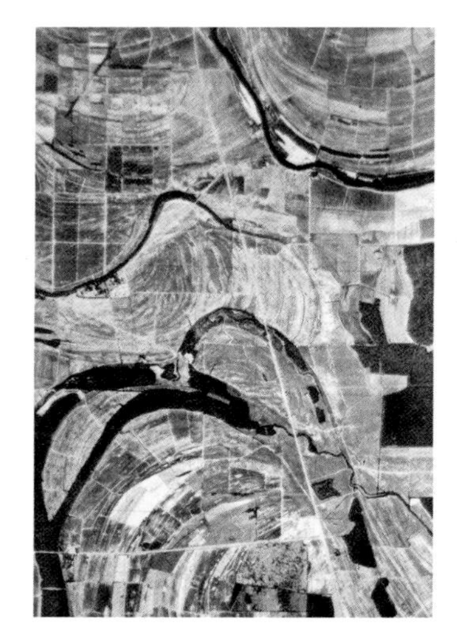

Aerial photograph of an area of the Yazoo Delta showing the order of agricultural fields superimposed on former meanders, from U.S. Geological Survey photoquad

Seen from space, the Yazoo Delta, like the rest of the alluvial valley, is a profusion of former meander belts, abandoned channels, and oxbows—the sweeping bends left by cutoffs that gradually silt up to become marshes, swamp forests, and, today, agricultural fields. This general evenhandedness of the Father of Waters in land-making over millennia was figured in a series of maps by Harold Fisk, a geologist at Louisiana State University in the 1940s. Today land-making has been stopped, perhaps momentarily. The present meander belt of the Mississippi, said to be twenty-eight hundred years old, has been fixed by engineers for the sake of the millions of people who live on it and in the drained and leveled backswamps like the Yazoo Delta.

Indian mounds as refuges during the 1927 flood
[Top: American Red Cross
Bottom: Bettman Corbis]

The land that the Mississippi has made in the delta is rich. John Barry describes it as "gold the color of chocolate. . . . Elsewhere one measures the thickness of good top soil in inches. Here good lush soil measures tens of feet thick." Native Americans constructed mounds in this soil before it was drained for plantations by slaves 150 years ago. Indian mounds, some of them as large as four hundred feet square and fifty feet high and in use when De Soto came through the Yazoo Delta in 1541, are popularly linked to religious and civic ceremonies and to the privileges of priests and chiefs. Only indirectly are they connected to a way of negotiating the spread of Mississippi waters. Whatever the reason for their existence, mounds in a backswamp that accommodate the Mississippi's shifts are in considerable contrast to a land barricaded from a river by continuous earthen embankments. They are two very different ways of inhabiting this dynamic terrain.[7]

Once numbering more than a thousand in the Yazoo Delta alone, mounds were leveled in the 1960s and '70s to fill depressions and make vast level fields for mechanized farms. The "book is closed on the mound-builders," writes Tony Dunbar; the "Delta now belongs to the land-levelers." Archeologists, late on the scene, perhaps because the Yazoo Delta lacked the glamour of an Egypt or a Mesopotamia, were barely ahead of bulldozers. Most often they were too late. Ironically, mounds were highlighted in the public view by aerial photographs of the 1927 flood, when together with levees they provided the only haven from the Mississippi depositing yet another layer of sediment on its backswamp.[8]

Levees as refuges during the 1927 flood
[American Red Cross]

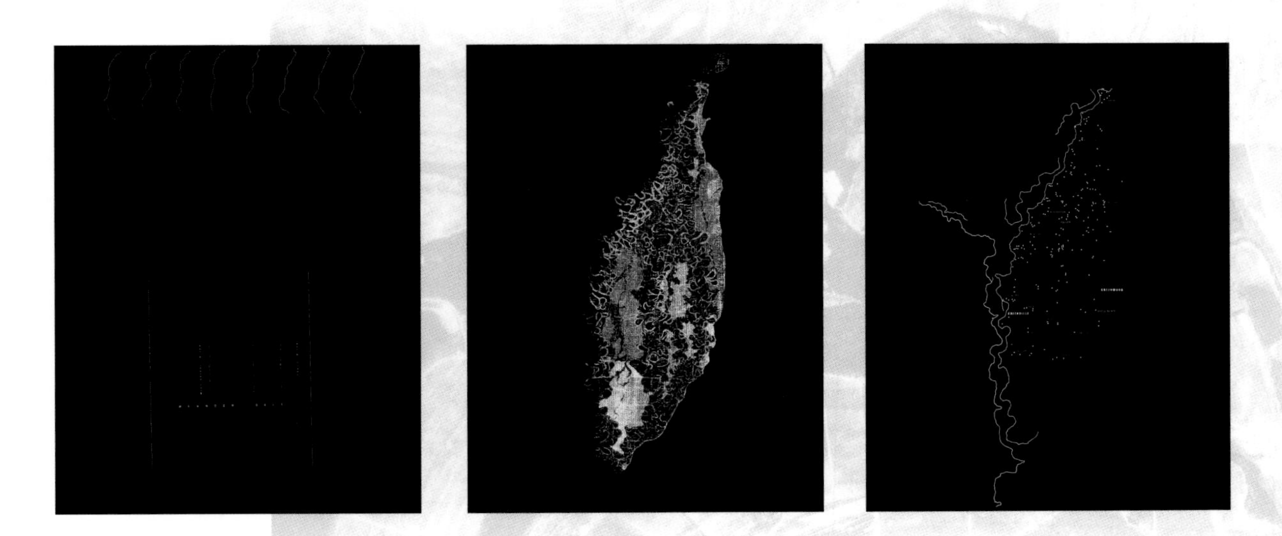

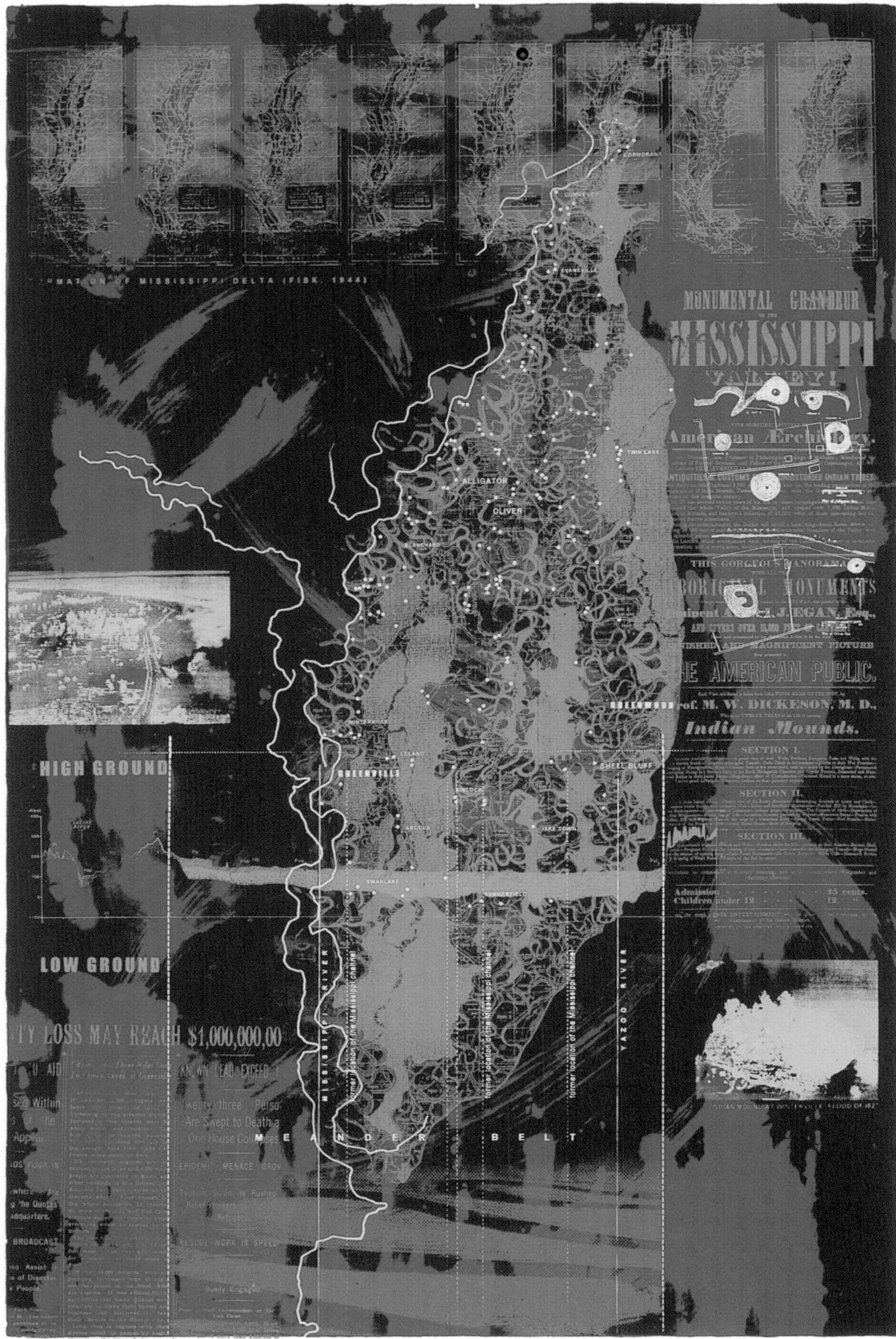

Depositing Depth (screen print on paper, 30" x 44")

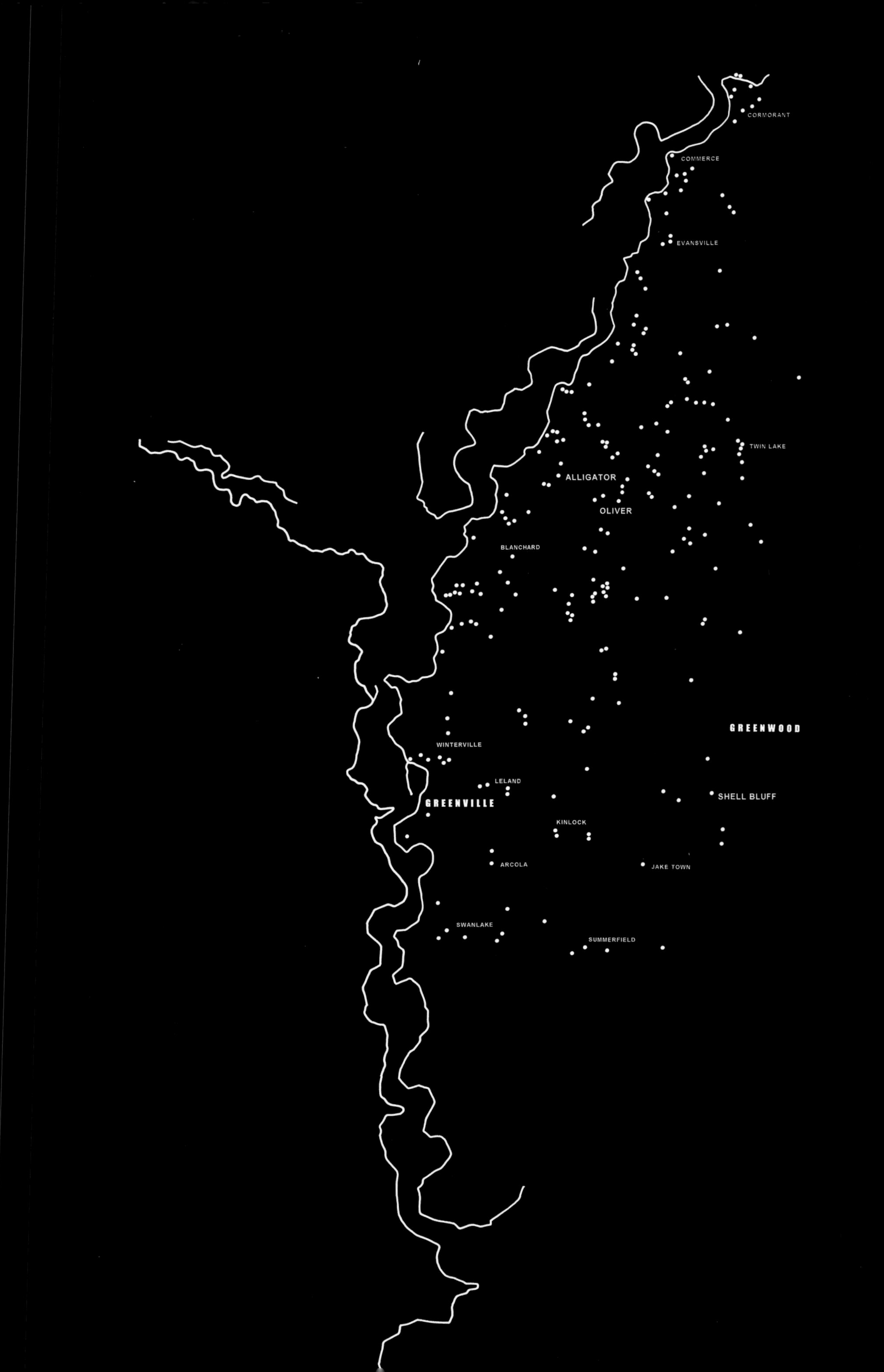
CORMORANT
COMMERCE
EVANSVILLE
TWIN LAKE
ALLIGATOR
OLIVER
BLANCHARD
GREENWOOD
WINTERVILLE
LELAND
GREENVILLE
SHELL BLUFF
KINLOCK
ARCOLA
JAKE TOWN
SWANLAKE
SUMMERFIELD

BLANCHARD
WINTERVILLE
LELAND
GREENVILLE
GREENWOOD
KINLOCK
ARCOLA
JAKE TOWN
SHELL BLUFF
SWANLAKE
SUMMERFIELD
GROUND
GROUND
MISSISSIPPI RIVER
former location of the Mississippi channel
former location of the Mississippi channel
former location of the Mississippi channel
YAZOO RIVER
MEANDER BELT
MAY REACH $1,000,000,00
KNOWN DEAD EXCEED
Admission
Children under

LATERAL MOVES

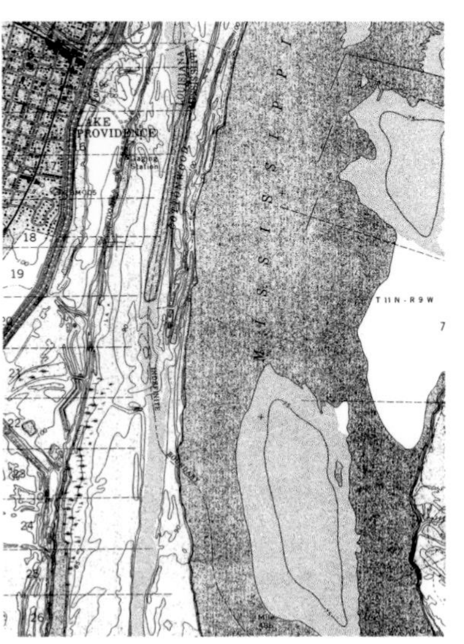

Indefinite boundary, from U.S. Geological Survey map, 1973

Despite evidence of the many former paths and abandoned channels of the Mississippi, a nation decided to rely on the "center line" of the river to mark its political boundaries. It is done, however, with some concession: the boundaries are declared "indefinite." The shifts of this "lawless stream" can, after all, as Mark Twain speculated, play "havoc with boundary lines and jurisdictions: for instance, a man is living in the state of Mississippi to-day, a cut-off occurs to-night, and to-morrow the man finds himself and his land over on the other side of the river, within the boundaries and subject to the laws of the state of Louisiana! Such a thing, happening in the upper river in the old times, could have transferred a slave from Missouri to Illinois and made a free man of him." In the case of the many cutoffs induced by the Corps of Engineers, however, such as those of the Greenville Bends in the 1940s, boundaries remain with the old route of the river, becoming definite even as the old channel becomes a still lake and eventually dry ground. Thus the State of Mississippi reaches across the river at Greenville to include small parcels of land enclosed by a definite boundary and a straightened river.[9]

Levees have kept the Mississippi from making dramatic shifts. Set back some distance from the water line, they do allow some room for lateral movement. Stone jetties protrude into the stream perpendicular to the flow, focusing the river's energy and velocity in a channel and away from its banks. Despite these efforts to keep the Mississippi focused on maintaining its position and moving sediments to the Gulf, some land is formed even as other land is destroyed. Islands surface here and disappear there. It could be seen as the Mississippi altering its course. But when the island is in someone's possession it is not allowed to disappear. Rather it is seen to move.

This is the case with Stack Island. In 1995 the U.S. Supreme Court decided that a sand bank that has extended the Louisiana side of the channel adjacent to the city of Providence over the past couple of decades is in fact Stack Island, which belonged to the State of Mississippi. The judge was convinced by a series of Corps of Engineers survey maps from over the years since the early 1800s. Overlaid, the maps made a case for an island that "keeps rolling along"—widening, diminishing, elongating, sliding, and finally attaching itself to the land mass of Louisiana, dragging with it the "indefinite boundary" between the contesting states to a new, grounded location. A report in the *New York Times* summarized the dilemma: "Were those 2,000 muddy acres—they are under water part of the year—the same muddy acres that were in the 19th century? Existential geology, anyone?" Engineered curves have to a large extent succeeded in keeping the Mississippi from moving, but they have hardly been able to keep it from moving boundaries—material and judicial.[10]

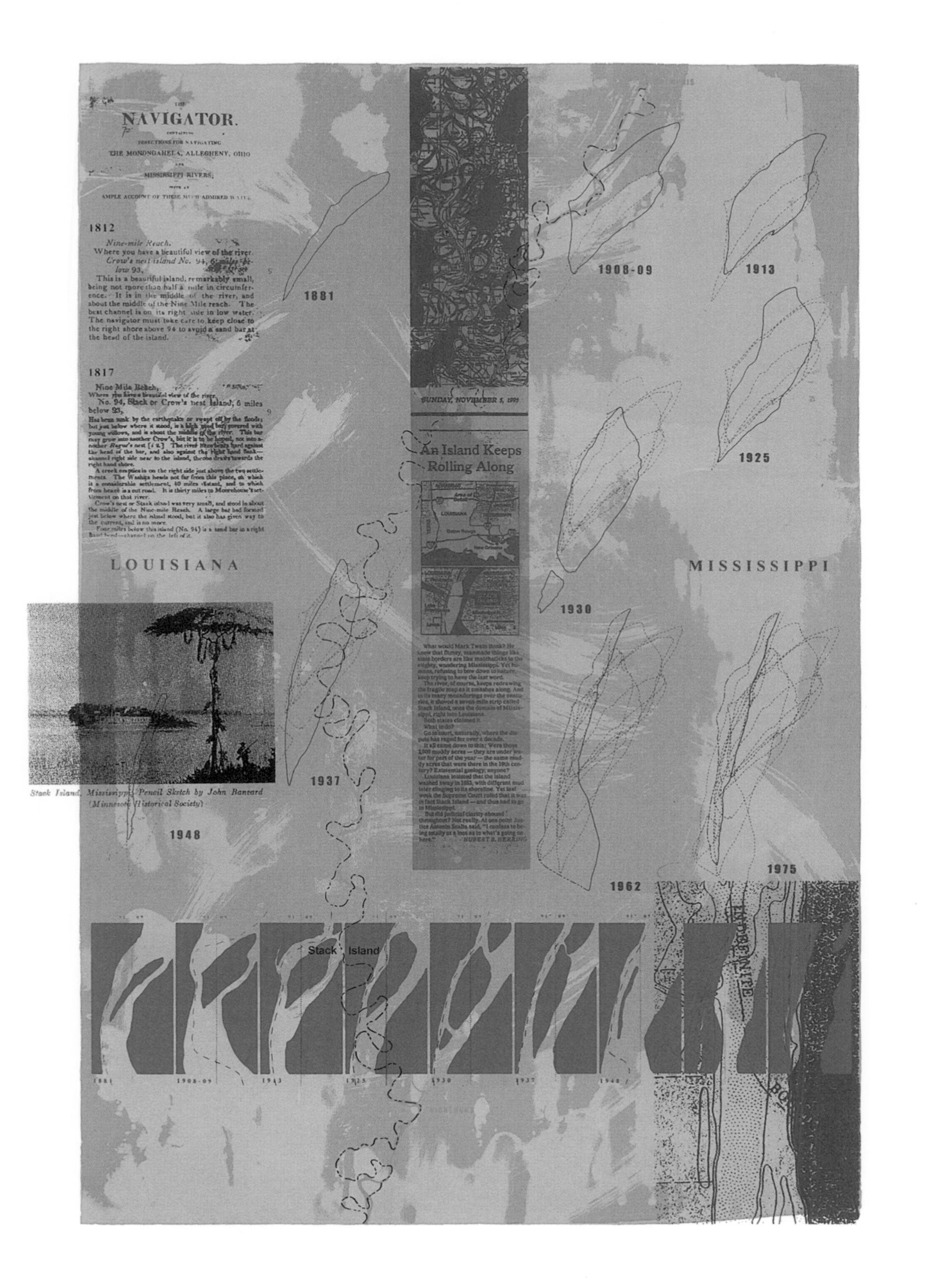

Lateral Moves (screen print on paper, 30" x 44")

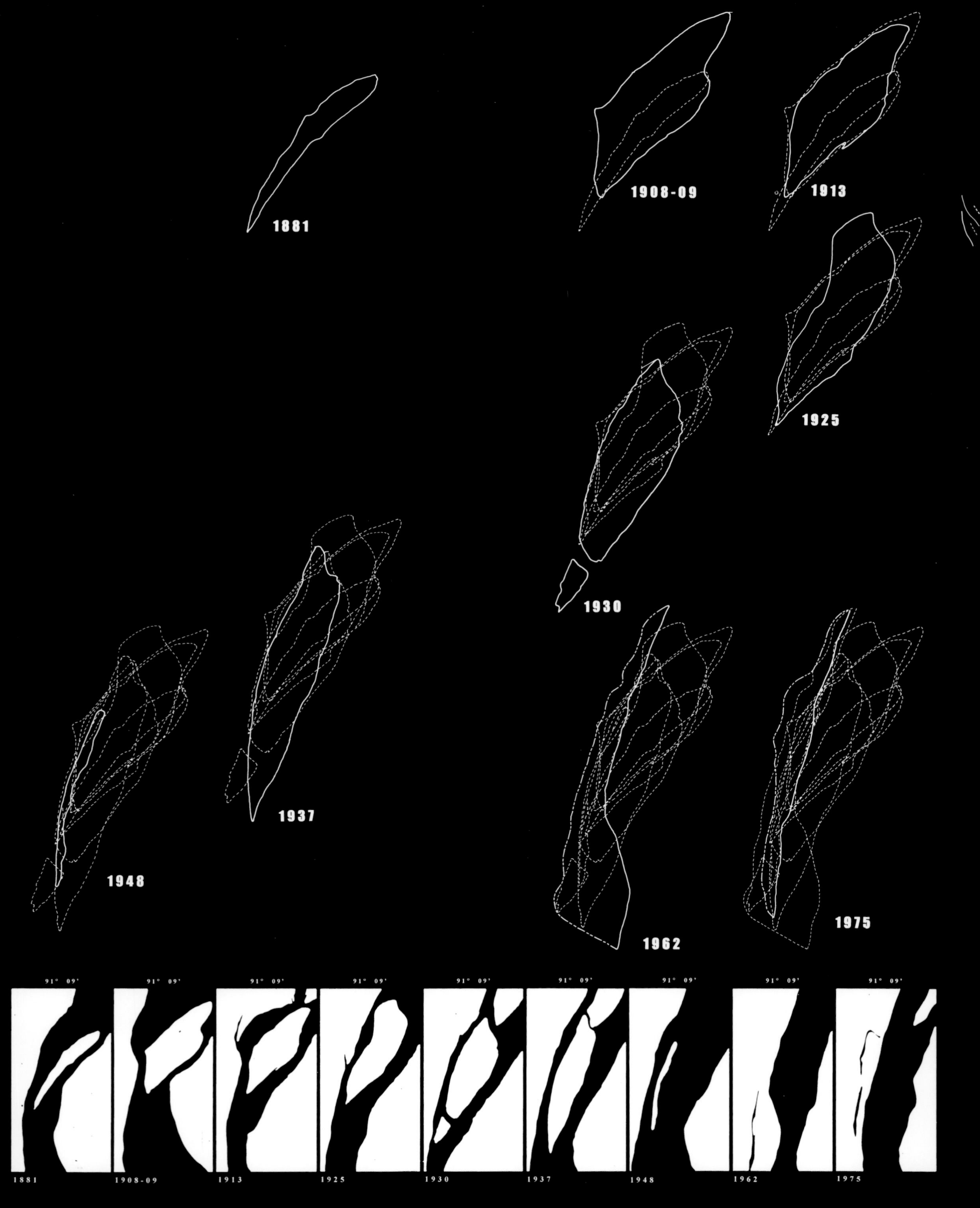
1881
1908-09
1913
1925
1930
1937
1948
1962
1975
91° 09'
1881
1908-09
1913
1925
1930
1937
1948
1962
1975

An Island Keeps Rolling Along

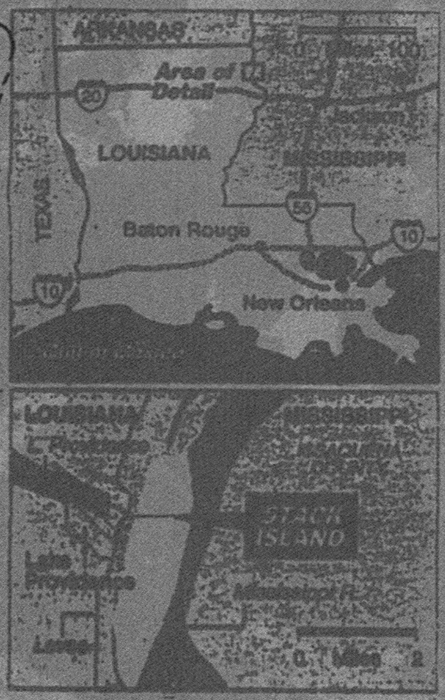

ARKANSAS
Area of Detail
LOUISIANA
MISSISSIPPI
TEXAS
Jackson
Baton Rouge
New Orleans
STACK ISLAND
Lake Providence
Mississippi R.

What would Mark Twain think? He knew that flimsy, manmade things like state borders are like matchsticks to the mighty, wandering Mississippi. Yet humans, refusing to bow down to nature, keep trying to have the last word.

The river, of course, keeps redrawing the fragile map as it smashes along. And in its many meanderings over the centuries, it shoved a seven-mile strip called Stack Island, once the domain of Mississippi, right into Louisiana.

Both states claimed it.

What to do?

Go to court, naturally, where the dispute has raged for over a decade.

It all came down to this: Were those 2,000 muddy acres — they are under water for part of the year — the same muddy acres that were there in the 19th century? Existential geology, anyone?

Louisiana insisted that the island washed away in 1883, with different mud later clinging to its shoreline. Yet last week the Supreme Court ruled that it was in fact Stack Island — and thus had to go to Mississippi.

But did judicial clarity abound throughout? Not really. At one point Justice Antonin Scalia said, "I confess to being totally at a loss as to what's going on here." *HUBERT B. HERRING*

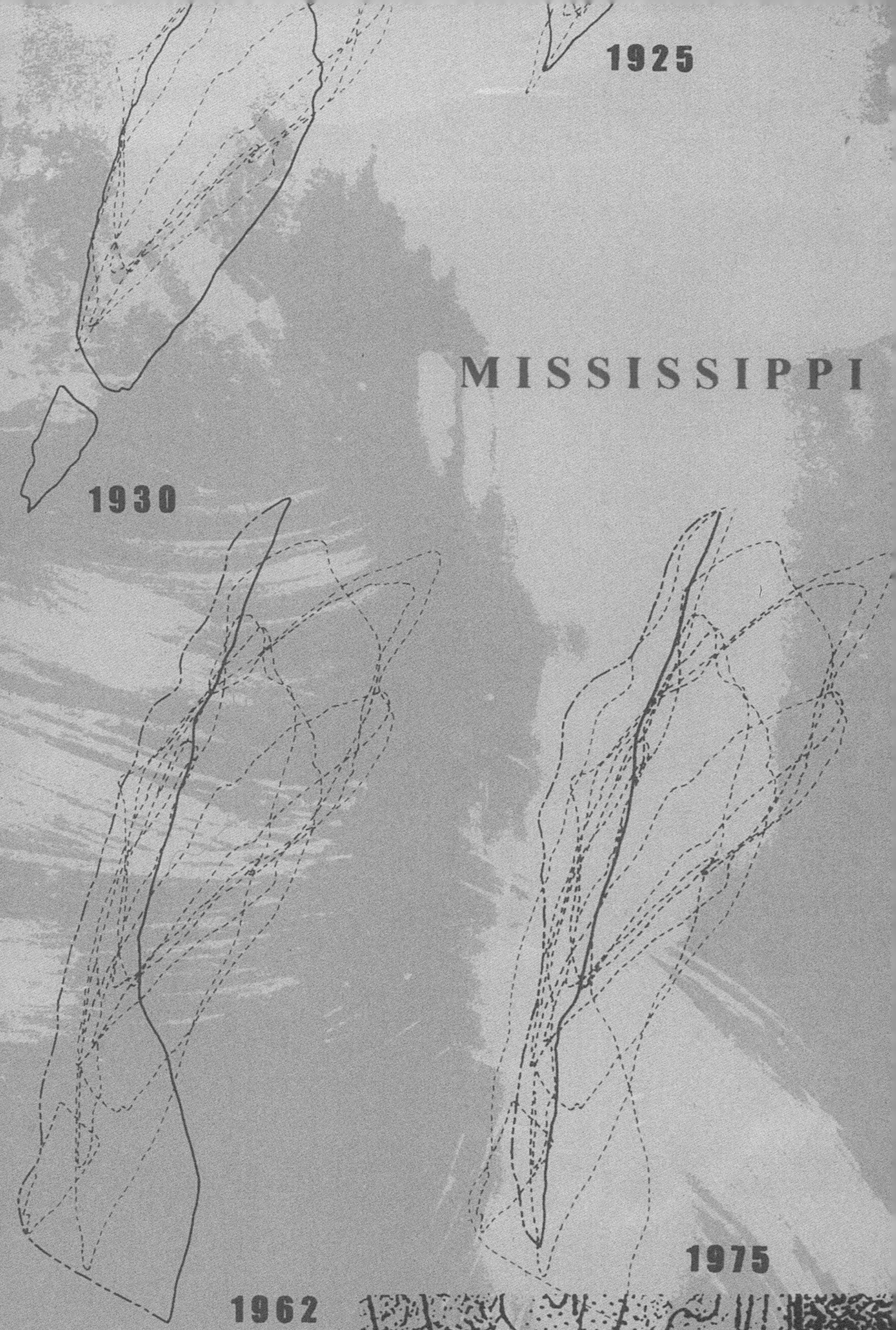

1925

MISSISSIPPI

1930

1975

1962

RAISING HOLLERS

"Suddenly, one raised such a sound as I never heard before, a long, loud, musical shout, rising and falling, and breaking into falsetto, his voice ringing through the woods in the clear, frosty night air, like a bugle call. As he finished, the melody was caught up by another, and then another, and then by several in chorus." Traveling the "cotton kingdom" of the American South before the Civil War and before he became the country's foremost landscape architect, the holler that Frederick Law Olmsted described was the musical cry of black slaves. As slaves, as sharecroppers when slavery was outlawed following the Civil War, as tenant farmers, and finally as day laborers in the service of corporations, black people brought forcibly from Africa have all along provided the muscle and music in the cultivation of the soil of the Yazoo Delta. It is acknowledged as the richest soil of the cotton kingdom. In the Yazoo Delta, more than anywhere else, blacks and soil are in the service of an "alluvial empire." 11

Cotton fields in the delta

Picking cotton in the Yazoo Delta in the early 1900s [Martin Dain Collection, Southern Media Archive, Center for the Study of Southern Culture, University of Mississippi]

A geologist in 1857 was already declaring: "It is still a wilderness... but after the lapse of another century, whatever the delta of the Nile may once have been, will only be a shadow of what the alluvial plain of the Mississippi will then be. It will be the central point—the garden spot of the North American continent—where wealth and prosperity culminate." After a century this was true for a minority. The forty-foot-thick topsoil laid by a meandering Mississippi—sandy loams mixed with "buckshot" clays—made wealth a reality for planters. The Yazoo Delta led the nation in the yield of cotton, thanks to the humid climate, short mild winters, and, above all, the "favorable

Bales of cotton along Route 1

dew moisture, vegetation, and soil conditions accumulated over countless centuries" that kept the boll weevil at bay (at least from the variety of cotton popular with growers there). *The Call of the Alluvial Empire*, a pamphlet designed to attract investment and labor in the latter half of the nineteenth century, declared that the cotton in the Yazoo Delta grew head-high, and experiments showed an acre to yield 220 bushels of corn—the second largest crop—as compared with the 40 considered excellent in the Midwest.[12]

In spite of its rich potential and its former wealth, the delta today is largely outside mainstream American life and prosperity. A recent traveler writes: "Without a doubt the Delta is economically one of the poorest spots in America. The tough life of the majority clashes with the apparent richness of soil and the great amounts of capital that are invested in farming it." Mechanized farming and scientific research for new cotton strains have taken over the laborious practices of cottonpickin'. Eight-row cultivators do the work of four men on plantations like the Delta Pine, once the world's

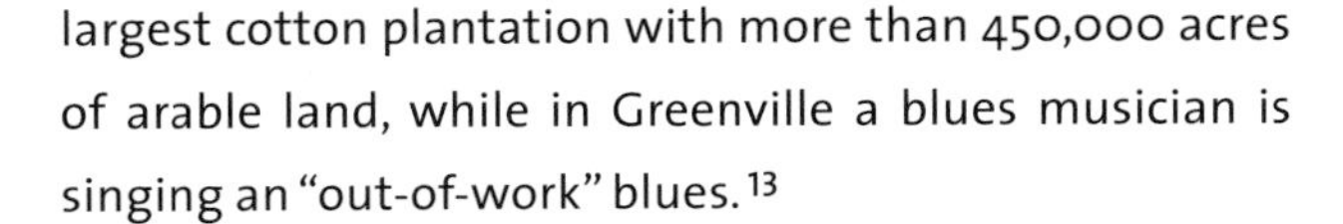

largest cotton plantation with more than 450,000 acres of arable land, while in Greenville a blues musician is singing an "out-of-work" blues.[13]

Blues bars in downtown Greenville

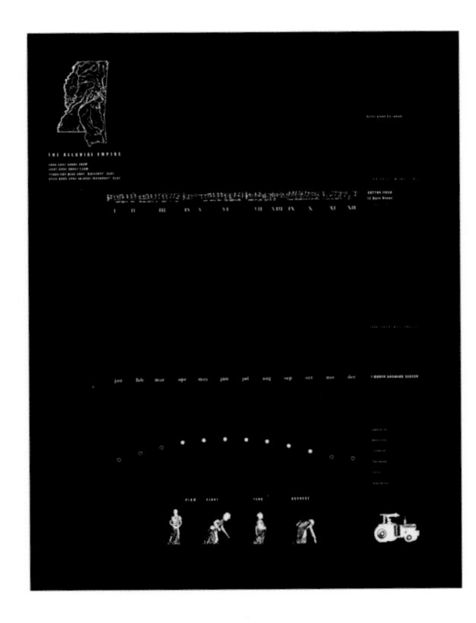

Raising Hollers (screen print on paper, 30" x 44")

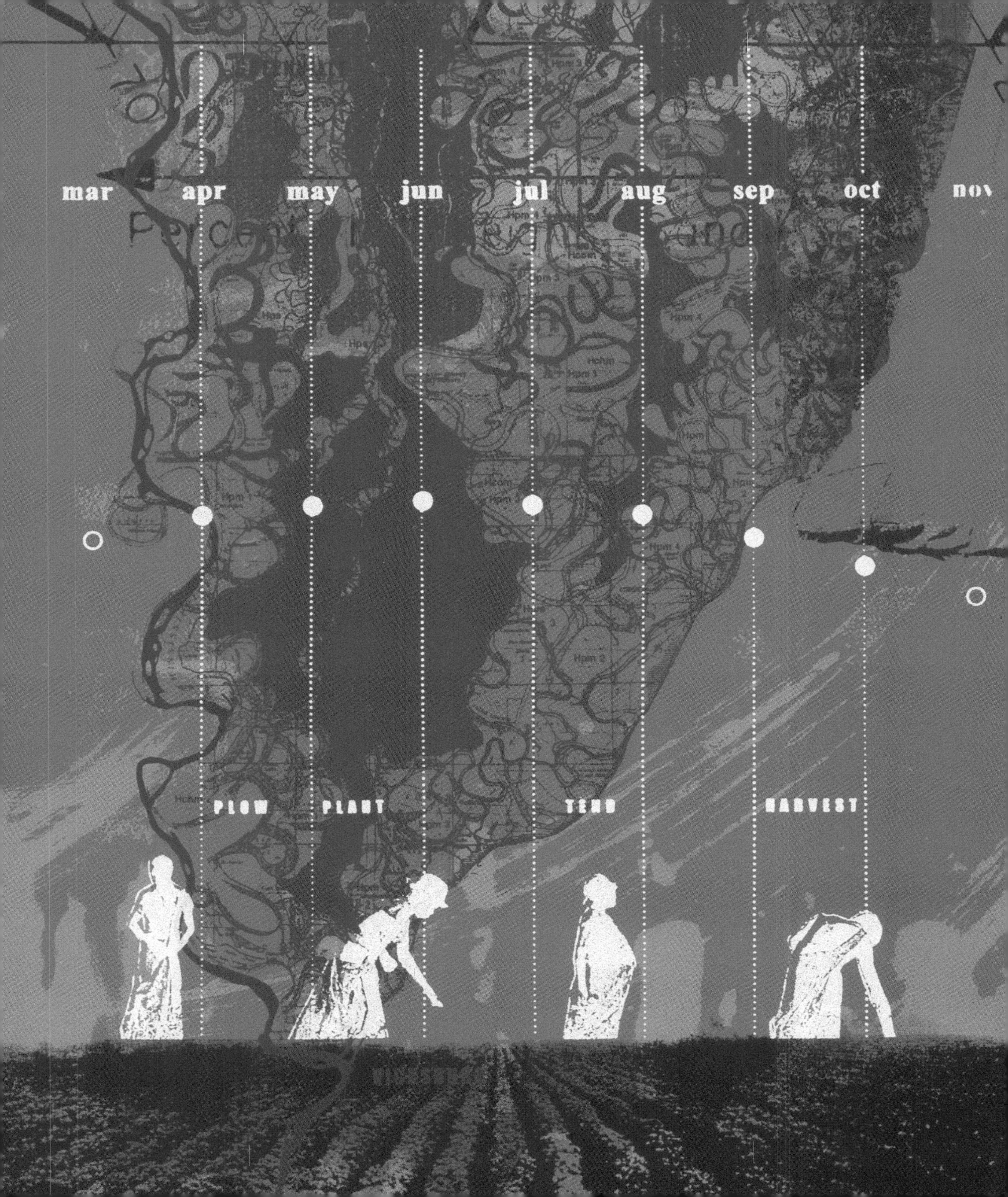
mar
apr
may
jun
jul
aug
sep
oct
nov
PLOW
PLANT
TEND
HARVEST

EXTENDING HORIZONS

On April 21, 1927, the Mississippi broke through the levee at Mounds Landing, 18 miles north of Greenville. The crevasse, three-quarters of a mile wide, gouged a channel one hundred feet deep for a mile into the Yazoo Delta. It was as if the Mississippi had decided to change course. The tremendous wall of water that first exploded through the crevasse and then continued to pour through it for weeks after inundated the lower half of the 7,000-square-mile Yazoo Delta. It spread across from Greenville to Greenwood not in a quick surge but "at a pace of some fourteen miles a day.... Everyone who saw the water and heard it had that image branded in their minds forever, for it had the eeriness of a full eclipse of the sun, unsettling, chilling." The silence and stillness that followed after the first thirty-six hours of mayhem, as half the Yazoo Delta lay under two to eighteen feet of water, was relieved partially by the unfamiliar drone of airplanes. The extended horizons of these relatively recent flying machines on reconnaissance and rescue missions matched the Mississippi's own. The river was 120 miles wide in parts.[14]

The vast inundated horizons of the Yazoo Delta seen from airplanes captured the public imagination. "For mile after mile all the land in view was the tops of the levees, to which thousands had fled for safety. In places the tops of giant cypress and oak trees still swayed in the breeze, the only green spots in the picture," in the description of a *New York Times* reporter flying over the fluid terrain of chocolate brown that dissolved all boundaries. Making daily front-page news in the national press from April to June 1927, images of the Mississippi in flood moved emotions and politics, making room once on May 22 for another aviation victory: Charles Lindbergh's historic flight from New York to Paris. In the long run, aviation contributed to a greater federal commitment and control of what comes across in testimonies and policy statements following the 1927 flood as an enemy that it was the national duty to defeat. Though the flood of 1927 was smaller in extent than the flood of 1882, its impact on a nation was far greater.[15]

The airplane and its much publicized horizons of flood, 1927
[Left: American Red Cross. Above: Clifton Adams, National Geographic Society, Image Collection. Right: U.S. Army Corps of Engineers.]

The Mississippi in its momentary release of 1927 precipitated the next chapter in the struggle for its control. The Flood Control Act of 1928 sought not only to strengthen existing controls but to change the paradigm of flood control. The problem was no longer just a river, but the entire river basin. The Mississippi Valley Committee (not to be confused with the Mississippi River Commission) was set up by the Public Works Administration to make a plan for the use and control of water in the entire

Mississippi drainage basin. Its report, submitted in 1934, began with architect Daniel Burnham's famous words: "Make no little plans. They have no magic to stir men's souls." The plans that followed were, indeed, big and comprehensive, and a flood of the same magnitude has not occurred on the Lower Mississippi since. But there is nothing to suggest that the war is over. The Mississippi, they say, allows memory to subside and confidence to rise before it strikes again. Its measure is flood. Prior to 1927 people lived by this measure, experiencing "time not simply in years, but in flood years—1858, 1862, 1867, 1882, 1884, 1890, . . ."[16]

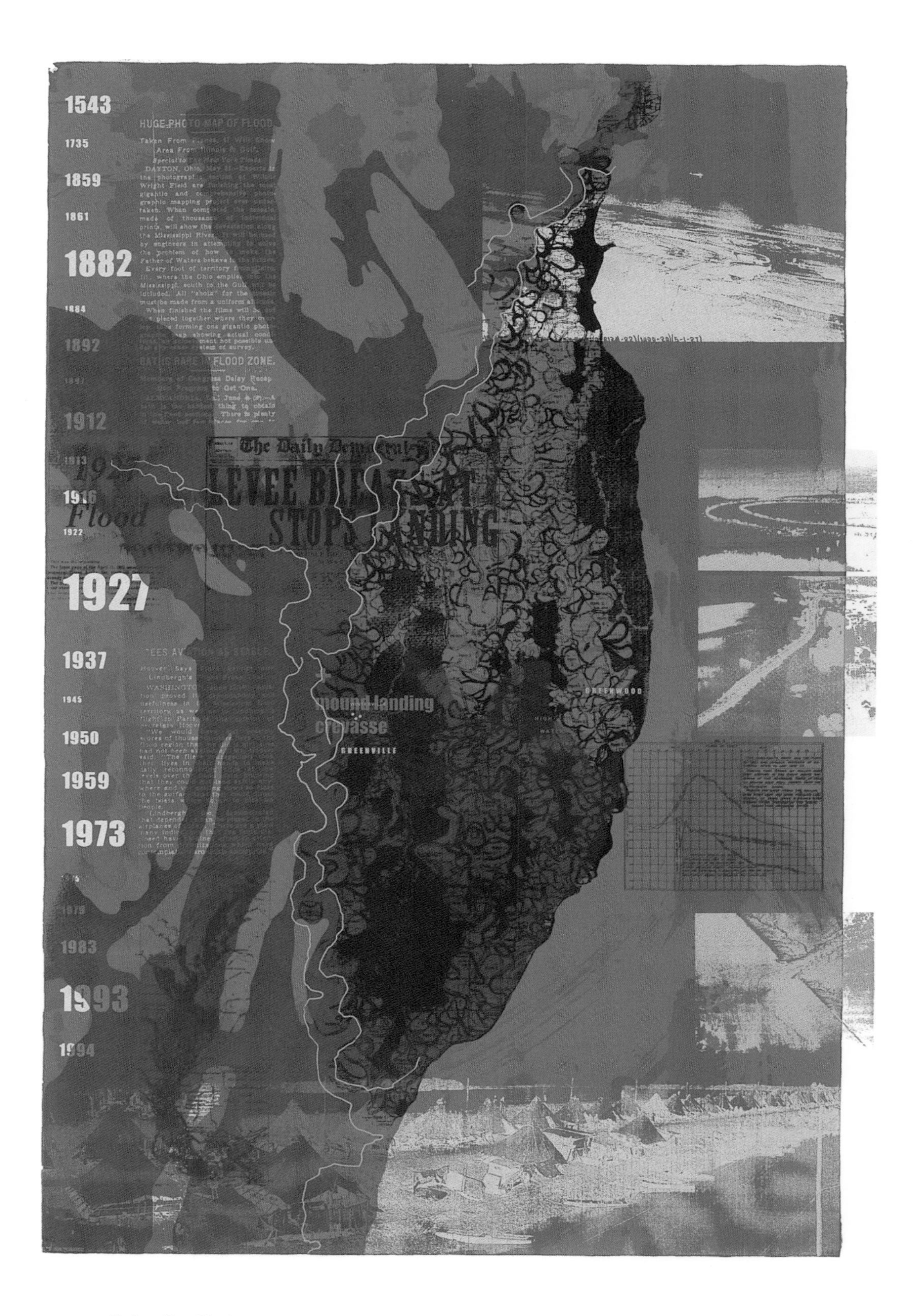

Extending Horizons (screen print on paper, 30" x 44")

DRAINING WATERS

The Yazoo Delta is more than just shaped like a leaf with its tip at Memphis and its stem at Vicksburg. Its surface is an intricate network of drains and rivulets including the outlets of four major reservoirs on the adjacent Mississippi uplands. The dendriform pattern of channels, natural and artificial, merges at one controlled point on the Yazoo River at Vicksburg, where the waters of this vast leveled plain are flushed out into the Mississippi on its way to the Gulf.

It took a planter's initiative and federal commitment to achieve this flushing. Andrew A. Humphreys, chief of the Corps of Engineers, referred to the Yazoo Delta in his famous 1861 survey of the Lower Mississippi as simply "that great Swamp." It was by many accounts "the wildest place on earth." What is today over 4.5 million acres of well-managed drained land, home to some 340,000 people, was virtually impenetrable hardwood forest and swamps when the first planters arrived here in the 1840s. According to author John Barry: "The density of growth suffocated, choked off air, held in moisture and a pulsing heat was so thick a horse and rider could not penetrate; even on foot one needed to cut one's way through. Only the trees, some one hundred feet high, burst above the choking vines and cane into the sunshine. Stinging flies, gnats, and mosquitoes swarmed around any visitor." Early planters sowed the seed of the alluvial empire on the natural levees of the Mississippi and its tributaries. These ridges, sloping into backswamps, were occupied by the Native Americans who had long since vanished, leaving behind the mounds that became building sites and yard features of plantation homes. But if the alluvium was to be fully exploited, the backswamps had to also be cleared, drained, and leveled.[17]

Clearing the Yazoo Delta, from *Washington County Drainage Bulletin*, April 1918

Draining the Yazoo Delta could not be a one-time event. In a humid subtropical land that receives heavy rains in the springtime when the Father of Waters rises with the melting snow from the fringes of its basin, draining is an ongoing challenge. At the turn of the century, drainage districts were established as the operational units to construct and maintain ditches, the lifeline of the new plantation society. "Our ditches are the highways for our drainage waters and we must observe the same rules for taking care of them as we do for our highways," urged a 1918 drainage bulletin of Black Bayou, Washington County. Axemen, moving just ahead of surveyors,

Rice fields in the delta

cleared lines through forests to make way for an assortment of walking dredges, floating dredges, and drag lines. Together they constructed a network of new and old watercourses, combining the meandering form of creeks and bayous with the rectangular ordering of fields. Work continued in small measure until the 1960s and '70s when, in an acceleration, any arable land that could be claimed by bulldozers was cleared, depressions were filled, and mounds demolished until each field was "perfectly, measurably flat." The vastness of the delta came into view.[18]

Today, water for the drained and leveled fields of the Yazoo Delta is pumped up from aquifers fifty feet below its surface. Gigantic irrigators, spanning more than a thousand feet from their point of connection to a well, draw mists of moisture over fields of soybeans, rice, corn, and cotton.

The extended terrain of mechanized farming

Pivot irrigators

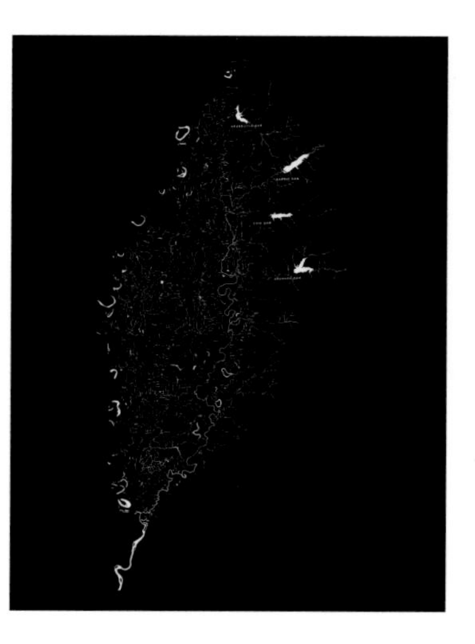

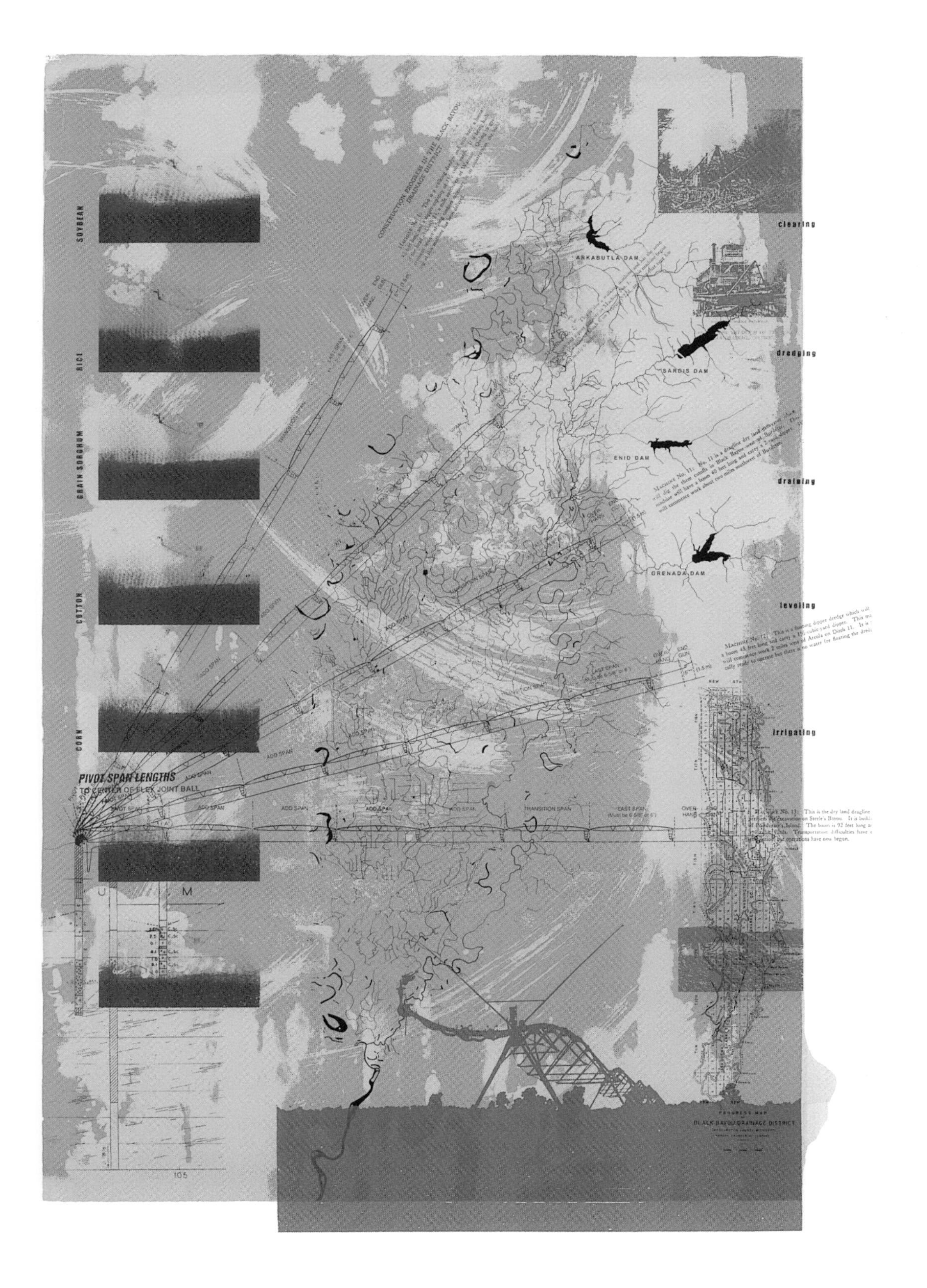

Draining Waters (screen print on paper, 30" x 44")

FLOODING SOIL

Although the stem of the leaf at the bottom of the Yazoo Delta is only an outlet today, there was a time in the recent past when it functioned as an inlet as well. High waters of the Mississippi backed up tributaries like the Yazoo, which in turn would back up its own tributaries, the Big Sunflower, Deer Creek, and Steele Bayou, and so on, spreading flood. Today this back distribution is disallowed by controls. In its place, however, is emerging from the heart of the Yazoo Delta a new ordering of flood, a progression of inundated fields bounded by low levees. Catfish is the new crop of the Yazoo Delta, and by most accounts a lucrative and self-reliant one. Unlike cotton, which was exported from the Yazoo Delta in raw form, catfish is bred, processed, and canned in the delta. Catfish ponds are spreading slowly. It will take more than three hundred years at present rates to cover the ground that the Mississippi did in a few days in April 1927. The methodical spread of this flooding is certainly in contrast to that Great Flood. These orderly waters sit in scientifically defined rectangular basins that shimmer like window panes in the Yazoo Delta, monitored for oxygen, pH values, ammonia, nitrites, hardness, and unwanted scale-fish, their beds free of all roots, stumps, and debris that may interfere with seining a harvest.

Low levees defining catfish ponds

Aerators at work in a catfish pond

Catfish ponds easily appropriate the geometry of cotton fields, their leveled surface and base of clay, and the elaborate drainage ditches constructed over a century. They average an efficient size of twenty acres and are generally built as four ponds around a well that pumps sixty million gallons of water for an initial fill, and forty million gallons a year thereafter, into each of them from the aquifer below. The six-foot-high levees that surround these shallow ponds provide access for feeding, harvesting, and careful monitoring, treatment, and pest control.

The catfish that Captain Marryat in 1837 found "the coarsest and most uneatable of fish" today is considered "perfect for the American palate," according to Tony Dunbar. They "take on whatever delightful flavor you add to them, bread them in, blacken them with, or fry into them." It is one of the most efficient meats for meal consumed. But it is also consuming the aquifer. Much of the water pumped up from the aquifer to feed the ponds is lost to evaporation and to a drainage system that served the overflows and the backwaters of the cotton kingdom so well. It was an imperial feat to clear, drain, level, and protect the Yazoo Delta—one of the toughest regions along the Big Muddy—from overflows and backwaters. With today's flood of increasingly scarce ground water, the memory of the vibrant swamps made by an abundant and meandering Mississippi has receded even further.[19]

Structured inundation—
the new horizon of
the delta

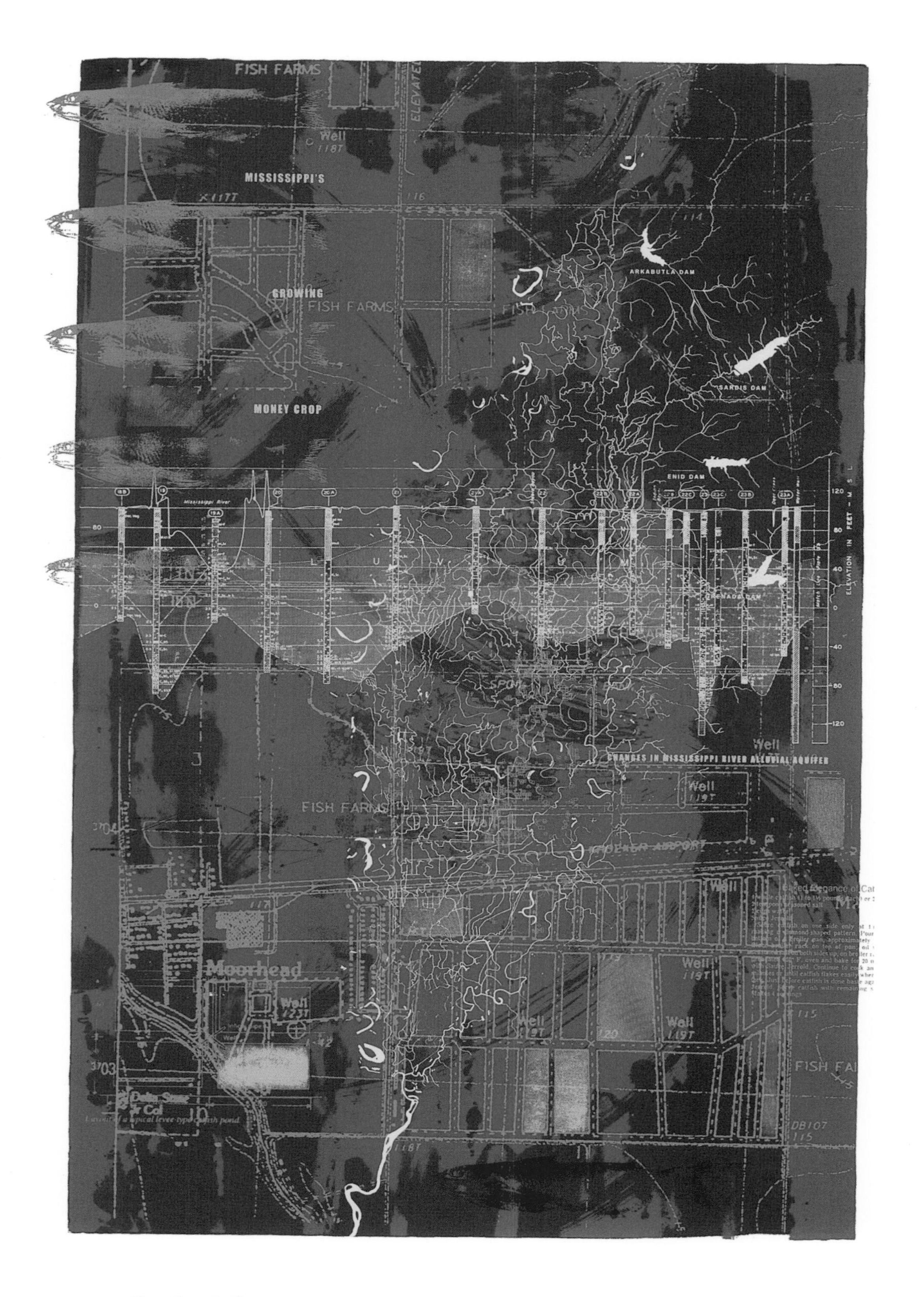

Flooding Soil (screen print on paper, 30" x 44")

GROWING

FISH FARMS

MONEY CROP

19

20

20A

21

21A

22

Mississippi River

19A

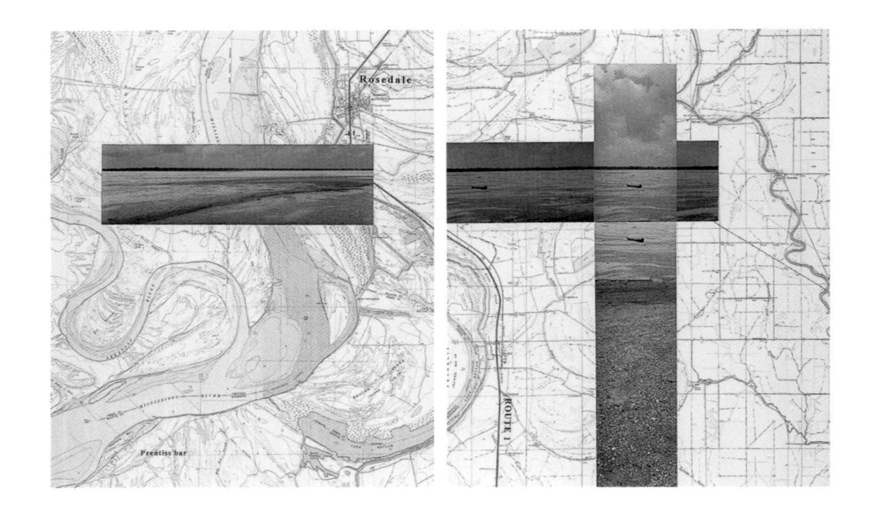

Across a Soil Trough
(color photo prints on USGS maps, each panel 15" x 17")

Route 82 descends into the flatness and heavier air of the Yazoo Delta at Greenwood from the uplands of Mississippi state between bluffs covered with kudzu. It is a world of clicking pivot irrigators and spraying dinosaur frames, dust clouds of plows, the drone of Cessnas on pesticide fly-bys, swishing aerators in catfish tanks, and roads claimed by farm equipment. The eye is carried to the horizon by long lines of earth furrows, drainage channels, cotton, rice, and soybean crops, the abandoned Illinois Central, and Route 61. Streams and the names of famous blues musicians mark the towns, motels serve agribusiness, and department stores are fronted by asphalt fields. At Greenville Route 82 crosses the Mississippi River over the forty-foot-high levees that contain the channel. At other places the levees move apart to hold thickly vegetated private hunting grounds and sandbars over a mile wide.

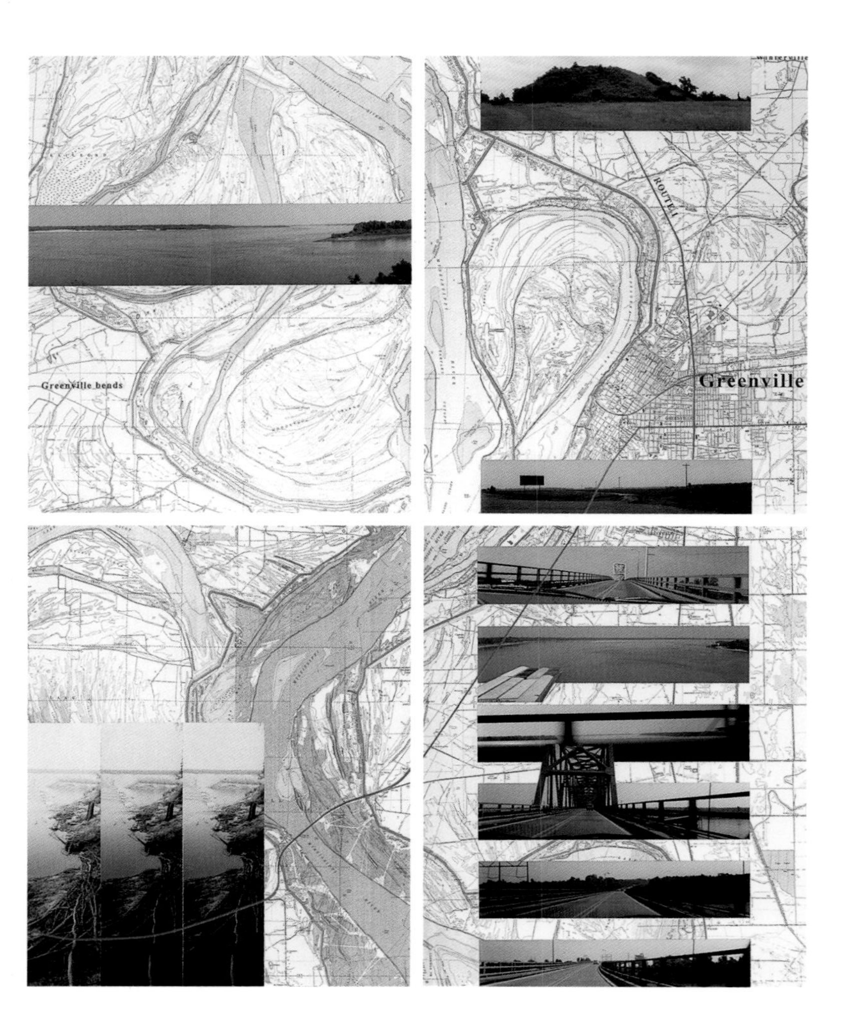

SITE 2 FLOWS

OLD RIVER CONTROL/ ATCHAFALAYA FLOODWAY

Liquid Earth
(charcoal on paper,
each panel 6" x 4")

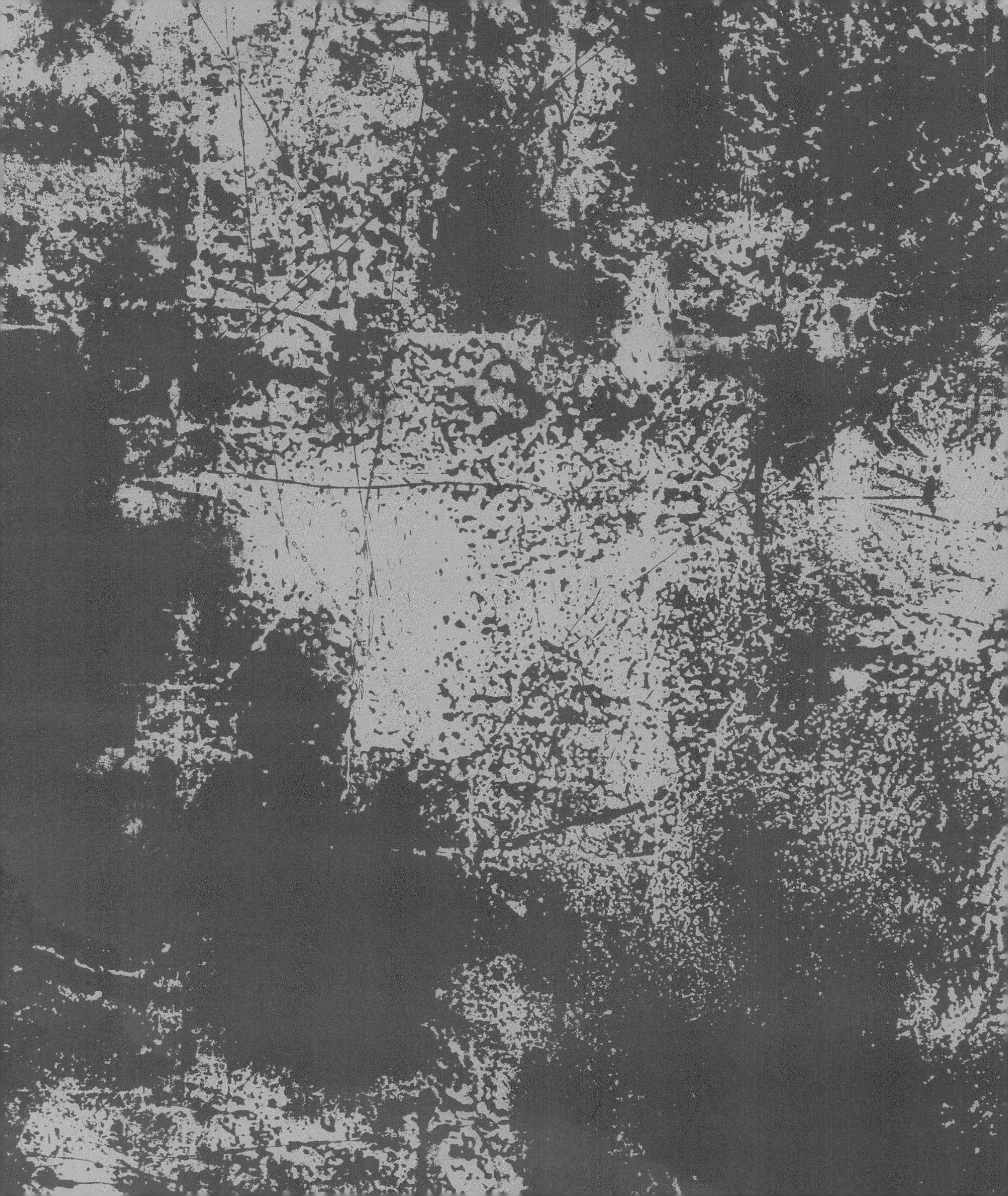

ARRESTING TIME

Site of the Old River Control Structures, Simmesport, Louisiana
[U.S. Army Corps of Engineers]

Geologists regard the Atchafalaya River as the first distributary of the Mississippi. The site of its departure from the Father of Waters at Simmesport, Louisiana, just below the mouth of the Red River coming in from New Mexico and immediately west of the Angola penal farm, infamous for its death-row inmates, marks a geological shift. It is the culmination of the alluvial valley and the beginning of the deltaic plain. But more than marking a geological shift, the Atchafalaya–Red River–Mississippi intersection is a site of profound design conflict. It is a conflict between the Mississippi River—which for millennia has swung like a pendulum from this point, forming the ground of southern Louisiana with the enormous amounts of its sediment—and the U.S. government, which is determined to hold it in place.

Construction of the Old River Control Structures
[U.S. Army Corps of Engineers]

Where the present meander belt of the Mississippi turns southeast to become the spine and lifeline of America's busiest industrial corridor and harbor, the Atchafalaya flows almost directly south through the intricate network of bayous that feed America's largest wetland. This was once the outlet of the Red River, independent of the Mississippi, until a large loop of the Mississippi called Turnbull's Bend eroded, silted, and flooded its way west to make the Red River its last tributary, and the Atchafalaya its first distributary. In 1831 Turnbull's Bend was cut off, separating the rivers once again. The upper arm of the abandoned bend silted up while the lower was kept open for navigation and trade. Following the Great Flood of 1927, this connection, called Old River by local inhabitants, was seen as a crucial junction for redirecting Mississippi waters in flood. Such redirection would help relieve the battered and bruised levees of the Baton Rouge–New Orleans corridor.

A decade and a half later, however, the diversion turned out to have unintended consequences, as the Mississippi began to favor the Atchafalaya. From the point of their divergence, the Atchafalaya takes the shorter route to the Gulf of Mexico—shorter by half. For the Mississippi, the Atchafalaya channel provides an enticing alternate. Harold Fisk, a noted geologist and author of numerous studies for the Mississippi River Commission in the 1940s, figured that if the water flowing in the direction of Baton Rouge fell below 60 percent of the total flow of the Mississippi, the river's meander belt would irreversibly shift to the Atchafalaya, a situation he predicted would occur by 1965–75. His argument was that the river by that time would be too slow to carry sediment, which would eventually choke the channel. Based on his findings the United States Congress in 1950 decided that it was not going to accommodate the vagaries of these shifting flows and risk the collapse of the "American Ruhr." It declared that "the distribution of flow and sediment in the Mississippi and the Atchafalaya Rivers is now in desirable proportion and should be so maintained." Thus, on a vast plain appropriately shared by the Angola penal farm, the flows of three rivers that have shifted and realigned several times in the past four centuries have been, as John McPhee describes it, controlled in space and arrested in time.[1]

The Auxiliary Control Structure [U.S. Army Corps of Engineers]

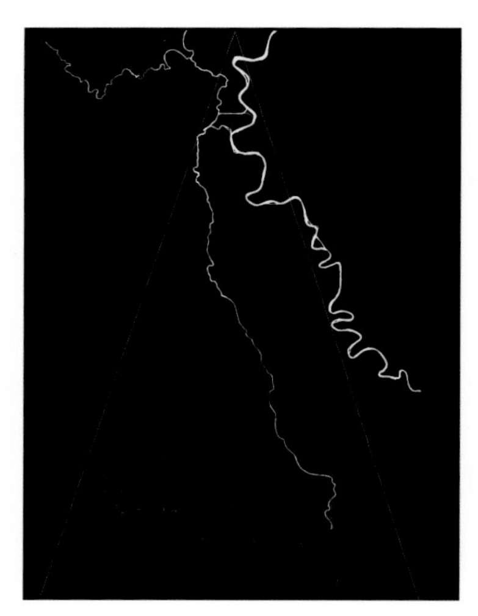

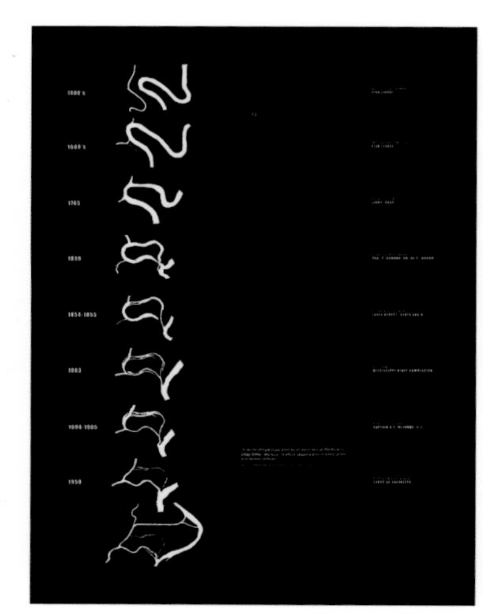

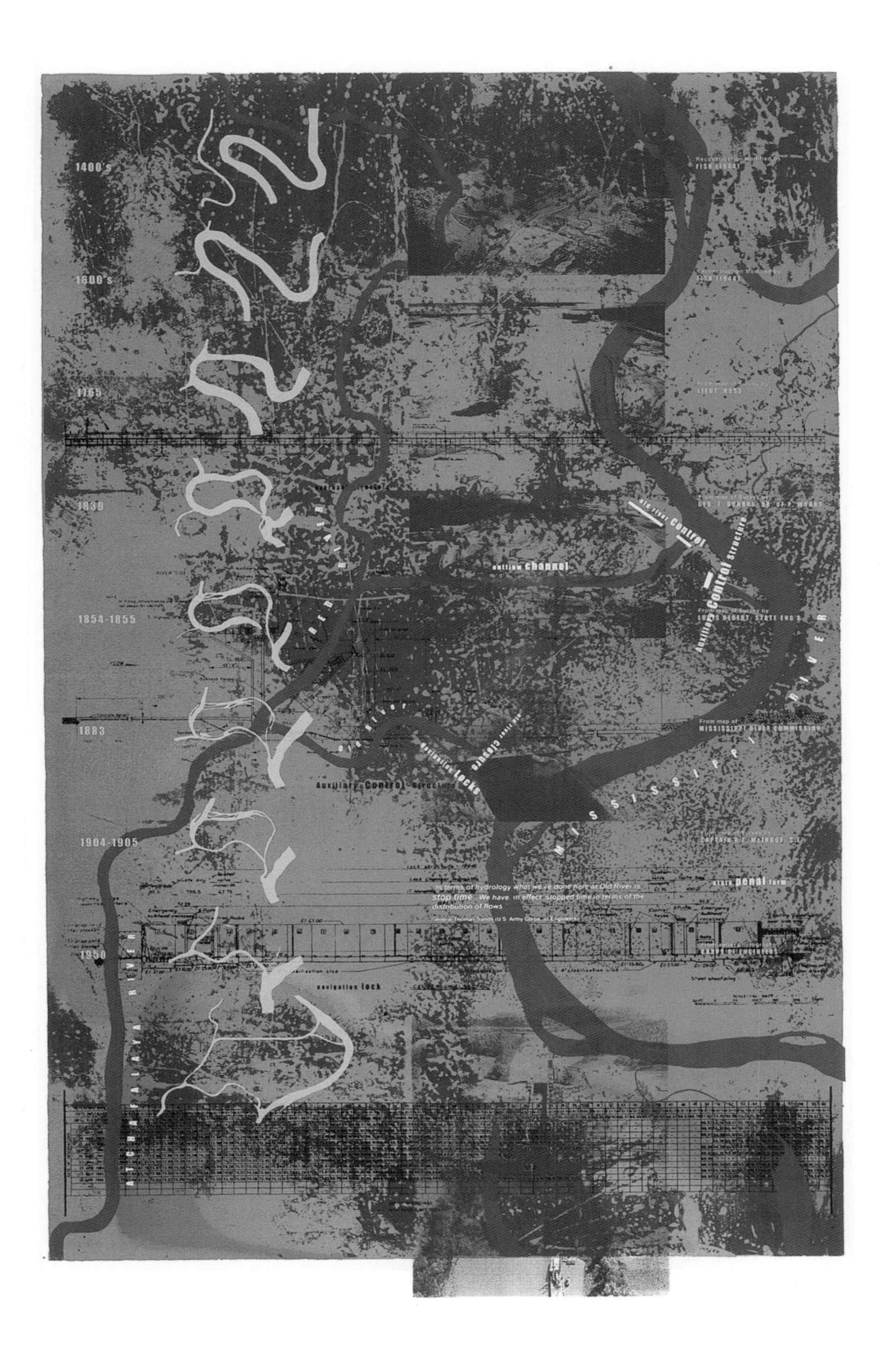

Arresting Time (screen print on paper, 30" x 44")

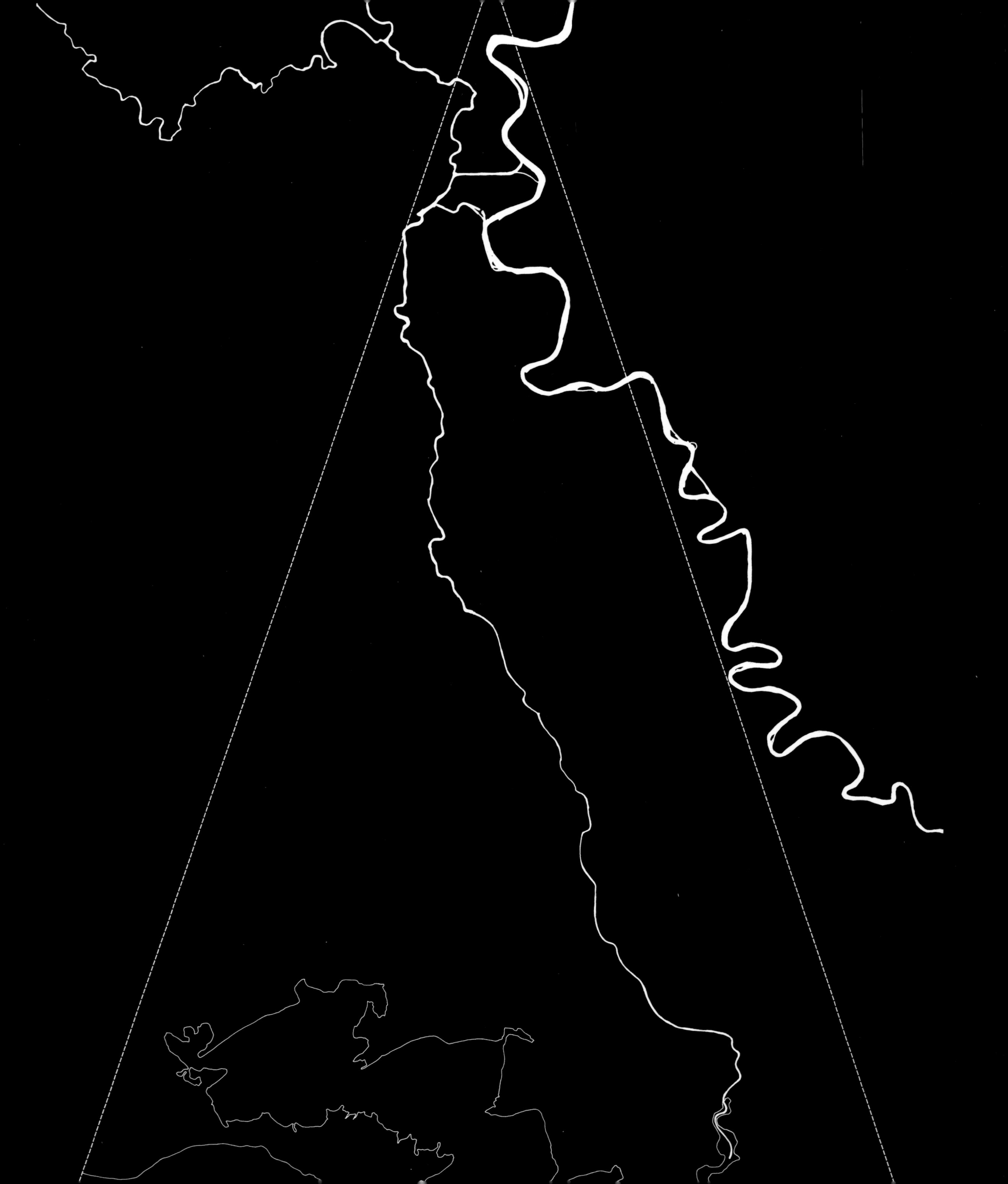

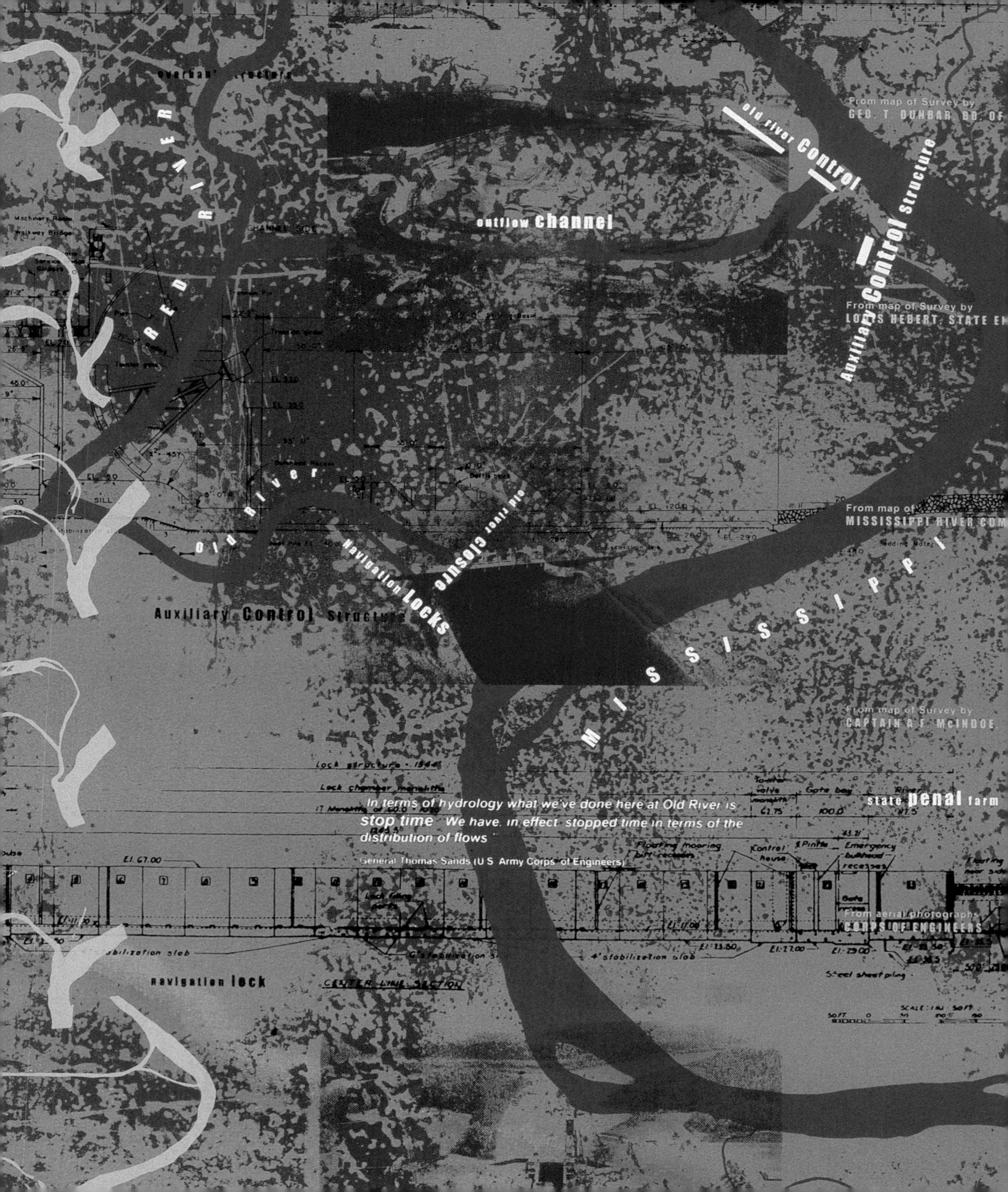

RED RIVER
outflow channel
Old River Control
Control Structure
Auxiliary Control
From map of Survey by
GEO. T. DUNBAR, BD. OF
From map of Survey by
From map of
MISSISSIPPI RIVER COM
Old River
Old River Closure
Navigation Locks
Auxiliary Control Structure
MISSISSIPPI
From map of Survey by
CAPTAIN A.F. McINDOE,
Lock structure
Lock chamber monolith
state penal farm
In terms of hydrology what we've done here at Old River is stop time. We have, in effect, stopped time in terms of the distribution of flows."
General Thomas Sands (U.S. Army Corps of Engineers)
Control house
Emergency bulkhead recesses
From aerial photographs
CORPS OF ENGINEERS
4' stabilization slab
navigation lock
CENTER LINE SECTION
Steel sheet piling

DIVIDING FLOODS

Where, in the worst-case scenario, the waters from an early hurricane from the east, late snowmelts in the Rockies, and torrential downpours in the south and central plains can arrive simultaneously, channeling waters through as many outlets as possible makes intuitive sense. It would save cities like New Orleans from being washed away. But with the Mississippi River Commission there was no empirical proof to validate this view that for decades had been advocated by a marginalized few. There was, on the other hand, sufficient evidence to back the thesis of the levees-only policy, that a river constricted and kept from spreading maintains a velocity high enough to push sediment and silt along while scouring its bed to deepen its channel. It took the devastation of the Great Flood of 1927 to allow the speculation on outlets to be given a chance.

Gates of the Auxiliary Control Structure

The Flood Control Act of 1928 acknowledged flood control of the Mississippi River as a national obligation. It promised a plan to protect against "the maximum flood predicted as probable" for the alluvial valley of the Mississippi River. Project Flood, as the plan was called, became the basis of flood-control works. It shifted the focus of the Corps from a river channel and its floodplain to the flows of the whole river basin.[2]

At the latitude of Old River the maximum probable flood of the Mississippi was set at three million cubic feet per second, an amount that would fill New Orleans with eight feet of water in six hours. Of this flow 30 percent was to pass into the Atchafalaya basin at Simmesport, and 70 percent was to continue down the

Mississippi. Two bends down, a further 20 percent was to be added to the Atchafalaya basin via the Morganza Floodway. Just above New Orleans another 10 percent was to be diverted into Lake Pontchartrain, leaving 40 percent to flow past a safe New Orleans. To manage the first step in this distribution the Old River Control Structure was begun in 1951. Designed as a sill, it was the new valve through which the water from the Mississippi fell into the Atchafalaya. Within a decade of its completion in 1963 it needed support, for it could not manage the floodwaters of 1973—the magnitude of which did not even come close to the estimate used as the basis of Project Flood. The Corps then built an Auxiliary Control Structure on a new channel to the south, which was completed in 1986. The two structures divide waters so that on one side of their gates are the calculated flows that keep the Mississippi in its present meander belt, feeding the prosperity of the Baton Rouge–New Orleans corridor. On the other side, sometimes as much as twenty feet lower, are "excess" flows that feed the labyrinth of bayous of the largest freshwater swamp in America.

Commemorative marker at the floodwall, Morgan City, Louisiana

Dividing the flows of the Mississippi on the basis of a hypothetical flood seems a simple proposition. But a Mississippi laden with tons of soil ever willing to settle and cause a flood makes the division hard to maintain. It doubles vigilance. Today the Corps' engineers must be as concerned for the inundation of Morgan City, near the mouth of the Atchafalaya, as they are for New Orleans, and as worried about the Atchafalaya channel as they are about the Mississippi below Simmesport. Perhaps Morgan City is in no more danger than if the Mississippi were free to choose its own ways. But a flood here is today tinged with professional agency. Project Flood has divided water, but it has also divided flood.

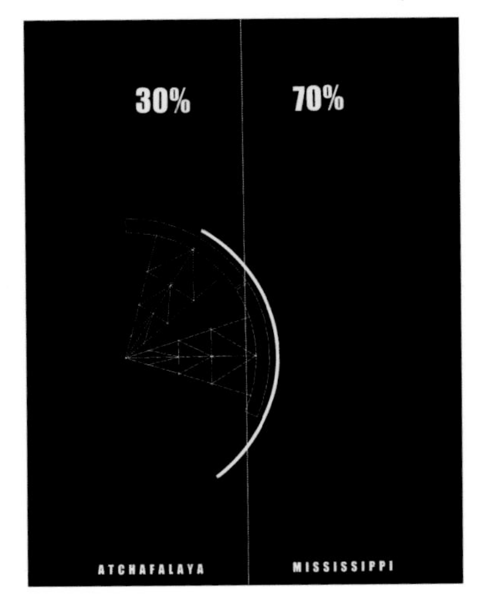

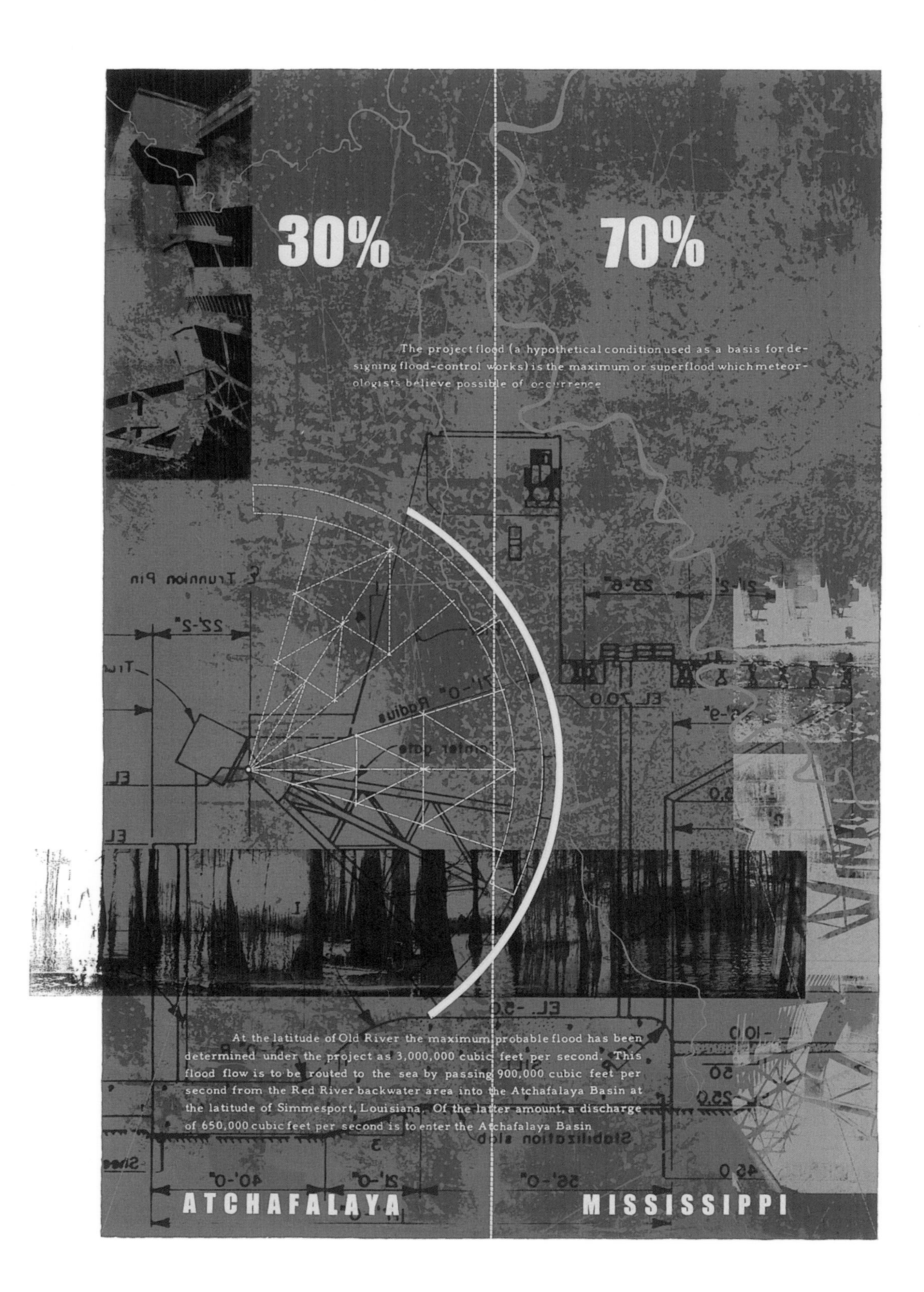

Dividing Floods (screen print on paper, 30" x 44")

SPREADING WATERS

The conflict between a river seeking its own levels and the Corps of Engineers determined to define those levels plays out not so much at Old River as it does in the "Cajun Triangle." Named for its most significant occupant, the Cajun, or *Cadien*, of French descent, the triangle is a world of bayou culture with its apex at the Old River Control and its base on the Gulf Coast. Here, the flow of labyrinthine bayous of the Atchafalaya basin, America's largest river-basin swamp, is in tension with the managed discharge of the Atchafalaya Floodway. Both spread waters, but they do so in different ways.[3]

The reflective waters of the Atchafalaya Swamp

The turbulent discharge at the Auxiliary Control Structure

The preference for a managed floodway over the complex flows of a basin, in many ways, precedes Project Flood. The ground of control necessary to making and managing an outlet was well laid by the interests of timber merchants who logged the giant swamp cypresses in the late 1800s, steamboat agents, and plantation owners. All were eager to have the water levels of the bayous controlled for navigation and transport to markets in New Orleans and elsewhere. Their interests have long coexisted in tension with those of Cajun farmers and fishermen whose subsistence way of life works with the rhythms of a basin. The ability of these farmers and fishermen to recede into the bayous in search of these rhythms has become increasingly difficult as the managing arm of the Corps of Engineers has extended. "The Corps," John McPhee observes, "has been conceded the almighty role of God." In large part this concession results from the fear of another flood of 1927, which devastated life in the triangle, threatening its status, not as the largest river-basin swamp, but as the sugarbowl of America.[4]

Basin protection levee of the Atchafalaya Floodway

Bounded by hills on the west and the Mississippi levees on the east and extending from Bayou des Glaises to Bayou Teche, the sugarbowl was inundated by a wall of water that came down from a crevasse in the levees of Arkansas in May 1927. The nation watched as the sugarbowl became an inland sea. The front page of the *New York Times* recorded the daily progress of waters, describing fifteen thousand Cajuns in flight from a "water avalanche more than 30 feet high." Project Flood and the Atchafalaya Floodway that followed were conceived with a clear view of the potent images of the path of these spreading waters.[5]

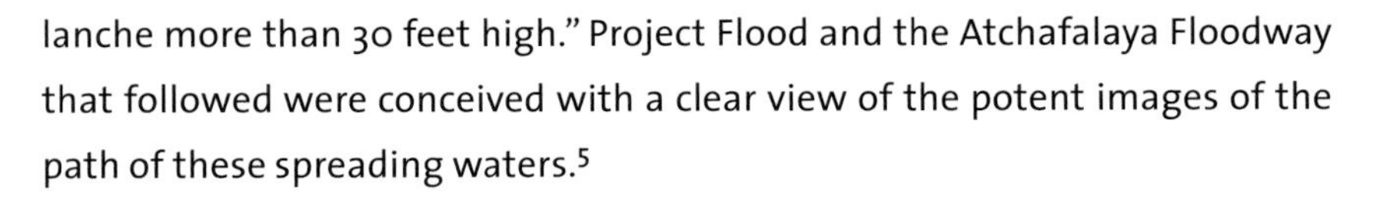

The floodway constructed by the Corps starting in the 1950s cuts a 15-mile swath through the Cajun Triangle for a length of approximately 140 miles with the Atchafalaya River as its spine. It actually comprises three floodways designed to carry half the combined waters of the Red and Mississippi rivers in the worst scenario—1.5 million cubic feet per second. The West Atchafalaya Floodway, with its source in a fuse-plug levee at the Old River junction, is the last resort. The Morganza Floodway begins two bends of the Mississippi below Old River. Both combine with the floodwaters carried by the Atchafalaya River itself, which enter deep into the basin between levees, to form the Lower Atchafalaya Floodway. The whole is bordered by 235 miles of protection levees.

Floodwall, Morgan City

At Morgan City, 10 miles or so above Atchafalaya Bay, the levees give way to 1.3 miles of concrete floodwall finished with bas-relief images of shrimp boats and oil rigs. The city sits precariously, like a "large tumbler glued to the bottom of an aquarium," its fate in the hands of the Old River Control Structures 140 miles above.[6]

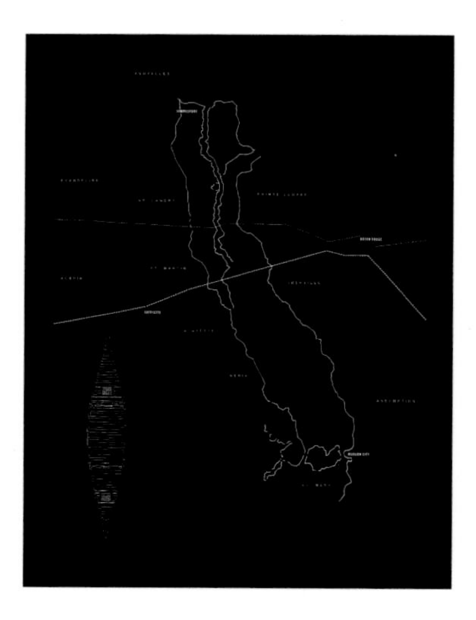

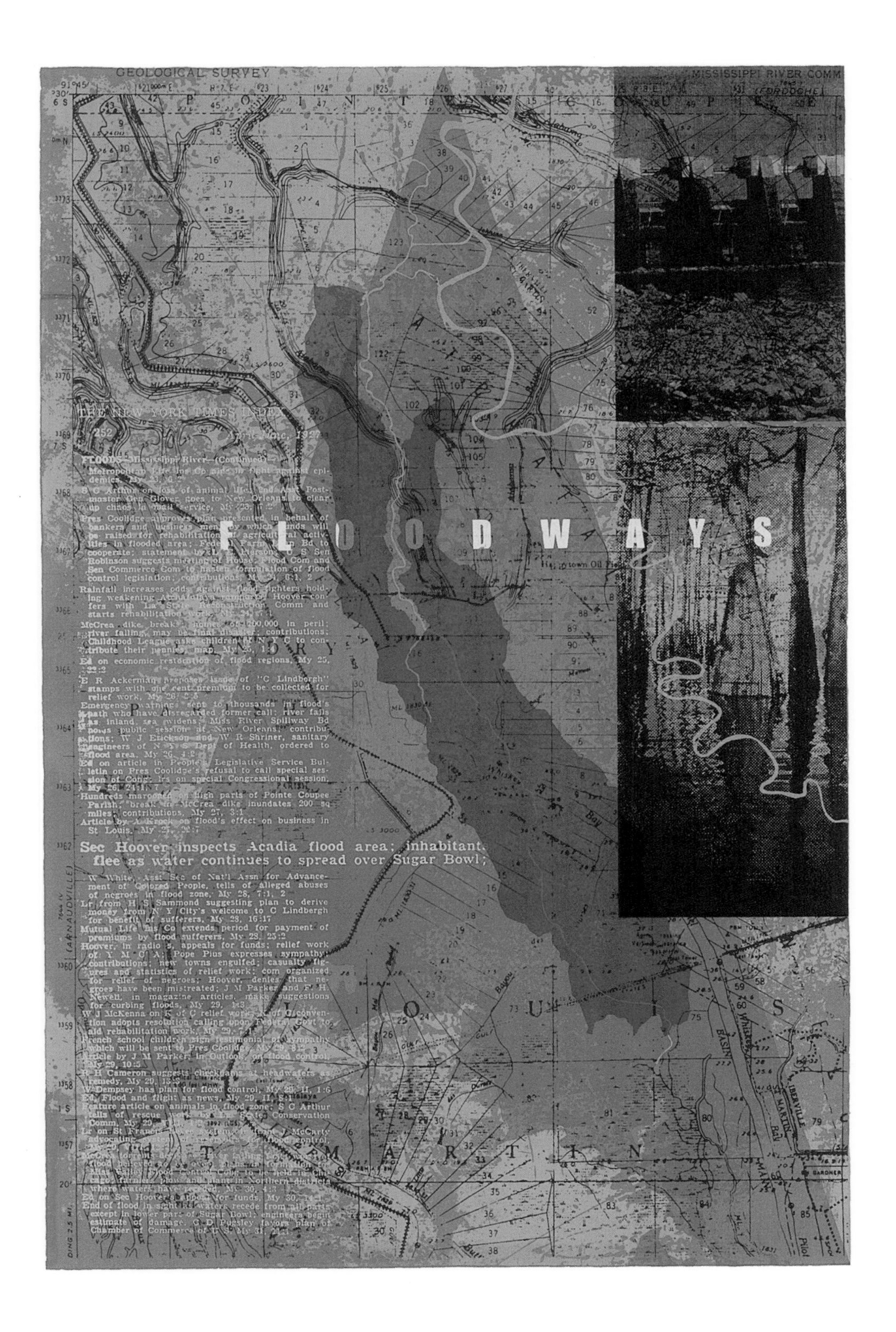

Spreading Waters (screen print on paper, 30" x 44")

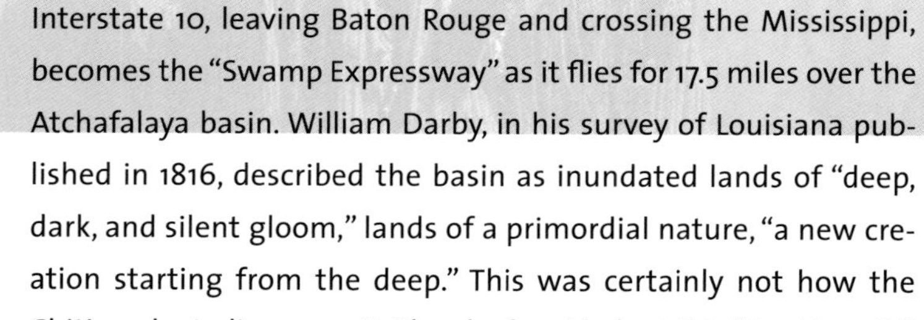

ACADIAN FLOWS

Interstate 10, leaving Baton Rouge and crossing the Mississippi, becomes the "Swamp Expressway" as it flies for 17.5 miles over the Atchafalaya basin. William Darby, in his survey of Louisiana published in 1816, described the basin as inundated lands of "deep, dark, and silent gloom," lands of a primordial nature, "a new creation starting from the deep." This was certainly not how the Chitimacha Indians saw it. They had resided on this "liquid earth" of swamps and marshes for centuries before being joined by French settlers and then Acadians, following their expulsion from Acadia (Nova Scotia today) by the English in 1755. Acadians began arriving in the bayous in 1765, when the French arpent measure was defining the European settlement pattern in Louisiana. The arpent is a French unit of length, approximately 192 feet. It was specifically applied to property as a measure of river frontage and lot depth along sides perpendicular to the path of the river. The result was a sequence of wedge-shaped properties along the winding bayous. They fan out from these waterways on the convex side, taper on the concave side, and negotiate in curious ways where bayous, more frequently than not in this labyrinthine world, flow in upon themselves. Here, while settlers from Virginia and the Carolinas cultivated sugarcane on plantations, particularly on Bayou Lafourche and Bayou Teche, both miniatures and former routes of the Mississippi, Cajuns fished and trapped in a basin teeming with life in Darby's silent gloom.[7]

Inhabitants and visitors in the lakes and bayous of the Atchafalaya Swamp

The Cajuns evolved from the Acadians, as distinct from the Creoles, whose ancestors are French immigrants who came directly to Louisiana from France beginning in the early 1700s. Whereas the Acadians had developed a New World identity, descendants of the French immigrants underwent a process of naturalization (together with descendants of immigrant slaves, animals, and plants) into an oppressive plantation culture on Louisiana soil called creolization. The subsistence ways of the Acadians kept them apart. They mixed with Native Americans and other Europeans, but they resisted the plantation regime, its Americanization following the sale of Louisiana to the United States in 1803, and its corporate identity today. Part of the Cajuns' culture is this evolving heritage of resistance, an Acadian flow, "willing constantly to reinvent and renegotiate their cultural affairs [but] on their own terms."[8]

The resistance of the Cajuns is not unlike that of the bayous, which defy another kind of ordering that, although not widespread in Darby's time, would perhaps have escaped him: dug canals and buried pipelines. The Atchafalaya basin is crisscrossed by hundreds of canals. At first they were dug to float out the virgin cypresses lumbered by "swampers" toward the end of the 1800s, as well as to transport goods and people. This is a terrain where, McPhee observes, "There is no terra firma. Nothing is solider than duckweed, resting on the water like green burlap." Today they serve as beds for oil and gas pipelines that run from fields dotting the basin to processing plants on the Mississippi between Baton Rouge and New Orleans. Revealed only by signs that warn against anchoring and dredging, the directed flows of these conduits, like the managed discharge of the Old River Control Structures above, contrast with the slow-moving bayous that keep this 1,250-square-mile swampland alive. The bayous move so imperceptibly that the French colonists who presumably adapted the word from the Indian *bayuk* (trans-lated as creek) referred to it as "sleeping water." But they also move mysteriously, changing direction almost unpredictably in this interconnected fluid terrain.[9]

Settlement on the bayou

The distinctive pattern of the arpent measure along Bayou Lafourche
[U.S. Army Corps of Engineers]

Acadian Flows (screen print on paper, 30" x 44")

On the bayou

LEVELING DISCOURSE

The Atchafalaya basin constitutes more than eight hundred thousand acres of a living, breathing terrain. At times the inhalations are so deep that water in many areas sinks below the surface of the soil. When it exhales there can be little dry ground. The variations are large, given the vast extent of the basin's sources. Creatures great and small, from crawfish to gar and alligators, depend on the cyclical water levels of this breathing giant. These creatures find their rhythms and tolerance levels met in different parts of the basin—cypress-tupelo swamps, hardwood forests, marshes, open lakes, and bayous. Given the shallow slope to the Gulf, the fragility of boundaries drawn by levels is evident. Economic activities like shrimping, crawfishing, trapping, and farming, as well as a range of recreational activities, share this fragility. They depend on water levels.

Since the 1950s the valves that control these levels have been in the hands of the Corps of Engineers. The presence of the Corps in the landscape brings forth the competing interests of the nearly three million inhabitants who depend on its management of flows. When crawfish farmers want more water, the shrimpers, who prefer brackish waters, demand less; the municipal interests, worried about salt water intrusions, and the navigation interests, concerned for dredging, appeal to maintain higher water levels; farmers upriver and downriver have a different perspective during high water on how fast water should be drained; oyster farmers oppose environmental interests that want to divert excess fresh water into the marshes; the Fish and Wildlife Agency and sportsmen join hands to oppose the Corps cutting off flows in the

Model of the Old River Control Structure at the Waterworks Experiment Station, Vicksburg
[U.S. Army Corps of Engineers]

many east–west distributary channels of the basin, which affect large portions of land that they both want to keep "wet and wild."[10]

More labyrinthine than the bayous are the lobbying voices of different interests. Today the Corps finds itself negotiating with an unending list of local, state, and federal agencies, no less complex than the bayous that first confronted them in the Atchafalaya basin. As part of the largest water resource controversy in the United States, the generals of the Corps set out every year in high and low water to hear from any group or individual who wants to stake a claim on the level of water in the Cajun Triangle.

Fishing and navigation, two of the many conflicting interests in the battles over levels in the Atchafalaya basin [Top: Louisiana Department of Wildlife and Fisheries]

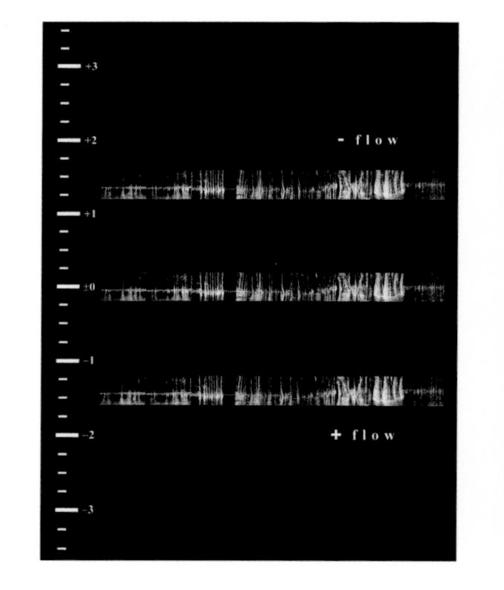

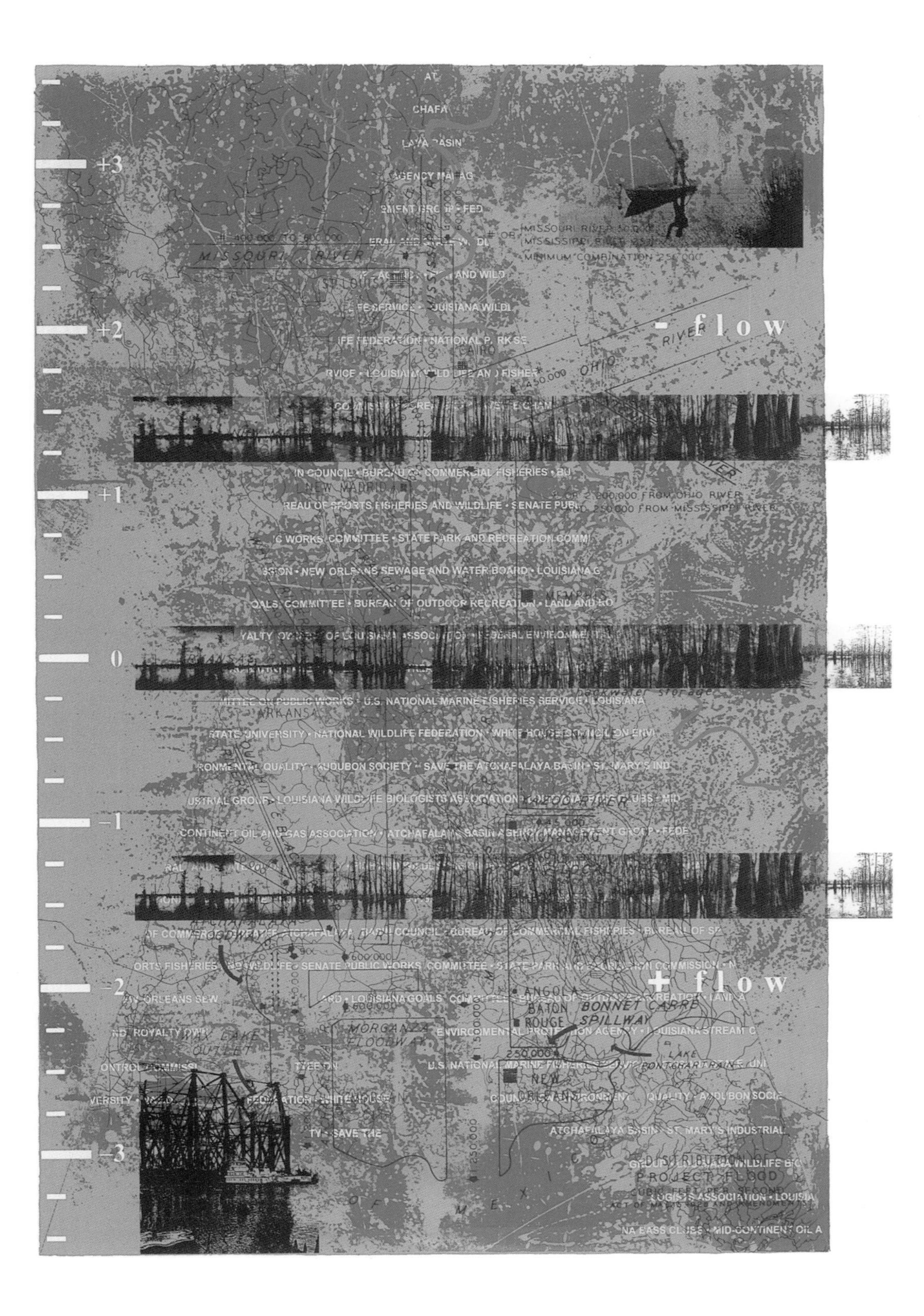

Leveling Discourse (screen print on paper, 30" x 44")

MOVING SEDIMENT

The Lower Mississippi valley is a land of moving sediment, where the moving is done both by the river and by the Corps of Engineers. The Father of Waters on a daily basis scours its bed, washes away banks, and changes its course, carrying in suspension tons of silt, trees, and assorted debris. Much of it gets to the Gulf, extending the continent, but some of it settles, and some of it is deposited on banks and backswamps in overflows, nourishing life and building land in an ambiguous act of construction and destruction.

The faster the river moves, the less the chance of overflow, but also the less chance that the sediment will settle before being carried out to sea. Increased velocity, then, solves the problems of flood and navigation. Where it does not work, moving sediment becomes an act of human labor. Or, rather, two acts—both of which have taxed inventive minds—dredging and snagging, the one to remove unwanted soil, the other to remove logs that can block sediments and vessels, especially when they collect in "rafts." One raft, building over the course of two centuries near the head of the Atchafalaya, stretched for forty miles downstream. Beginning in 1839 and continuing until 1860, the raft was burned and blasted to clear the Atchafalaya for trade to ports on the Red and Mississippi rivers. The clearing encouraged transport, but it also began the enticement of the River King that has led to the massive controls at Old River today.

With the constant vigilance in the Mississippi basin today, rafts can hardly be expected. But dredging is a daily operation, for soil is ever present in the Big Muddy, and settling soil is a perennial threat. But it is one thing to dredge the Mississippi, flowing between levees in an alluvial plain, and quite another to dredge the Atchafalaya, where levees end fifty miles into the basin and a braided system of bayous and lakes takes on the task of absorbing and distributing water and soil. The Gulf, through the labyrinth of the Lower Atchafalaya basin, can be very remote. In this network of flows, streams become lakes and lakes become streams in less than a decade, as land appears in one place and disappears in another.

Making this basin into a floodway following the 1927 flood did not require levees alone. It required a way to get floodwaters and sediment to the Gulf as quickly as possible, or else the wetland would be well on its way to becoming a hardwood forest. Thus, beginning in 1932, the Corps started dredging a channel through the open Lower Atchafalaya basin, from where the levees ended to Grand Lake, and ultimately to the Lower Atchafalaya River past Morgan City. In the two decades of intensive dredging that followed, the Corps displaced 127 million cubic yards of silt and closed a number of bayous with the spoils. The engineers soon found that their labor could not be measured by the depth of the river dredged, as would be the case elsewhere. It had to be measured by the volume of soil deposited, for land is made quickly in this basin. By 1950 sixty square miles of new land appeared in what once was the open water of Grand Lake and Six Mile Lake. It is a popular belief that the safest place to be in time of flood is now in the sedimented floodway. It is the highest land around.

An early dredge
[U.S. Army Corps of Engineers]

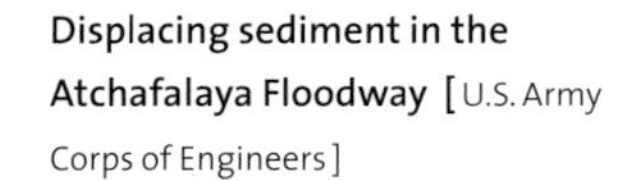

Displacing sediment in the Atchafalaya Floodway [U.S. Army Corps of Engineers]

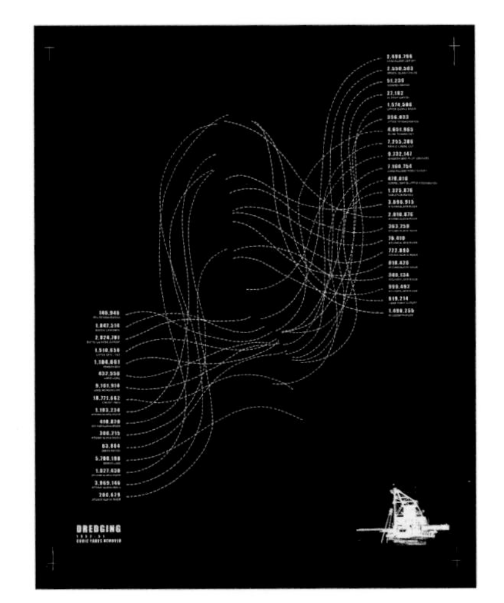

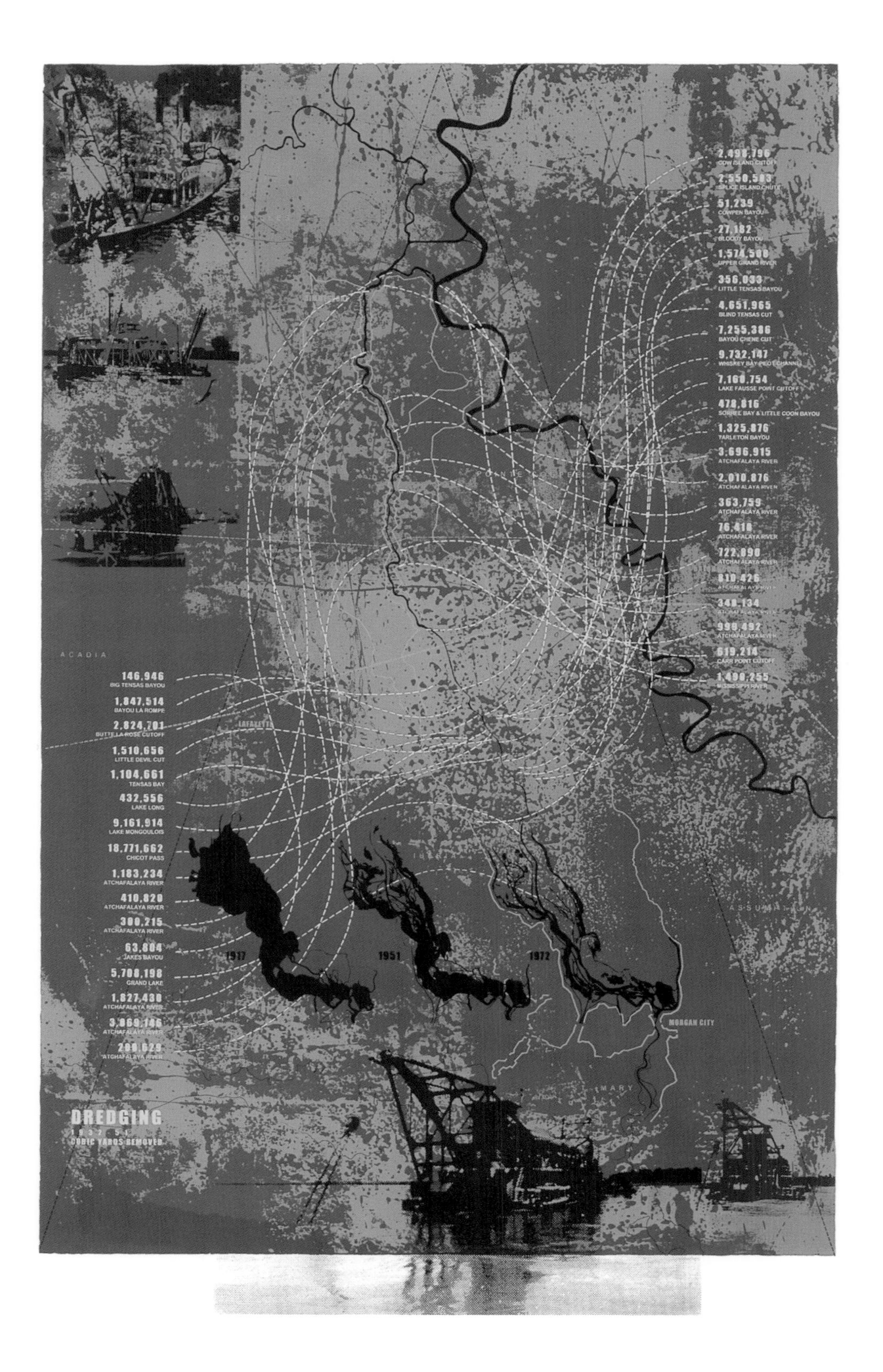

Moving Sediment (screen print on paper, 30" x 44")

1,847,514
BAYOU LA ROMPE

2,824,701
BUTTE LA ROSE CUTOFF

1,510,656
LITTLE DEVIL CUT

1,104,661
TENSAS BAY

432,556
LAKE LONG

9,161,914
LAKE MONGOULOIS

18,771,662
CHICOT PASS

1,183,234
ATCHAFALAYA RIVER

410,820
ATCHAFALAYA RIVER

300,215
ATCHAFALAYA RIVER

63,804
JAKES BAYOU

5,708,198
GRAND LAKE

1,827,430
ATCHAFALAYA RIVER

3,069,146
ATCHAFALAYA RIVER

200,629
ATCHAFALAYA RIVER

LAFAYETTE

LAFA

IBERIA

1917

1951

DREDGING

1932–51
CUBIC YARDS REMOVED

On Fluid Ground
(color photo prints on USGS maps,
each panel 15" x 17")

Ten miles from Baton Rouge, just across the East Levee, begins the Atchafalaya basin. Sluggish bayous, flowing sometimes one way and sometimes the other, clear or muddy, lead into and through a maze of cypress-tupelo swamps, marshes, open lakes, between thickets and cypress knees. Aboard a canoe in the deep swamps, there is no dry land to be seen, only liquid earth. The buttressed trunks of the bald cypresses calibrate the changing levels of this terrain. Here in the serene home of cormorants, fish, owls, nutria, and a host of amphibious creatures there is an awareness that change, perhaps dramatic change, can be effected by a switch at the Old River Control Structures, fifty miles north.

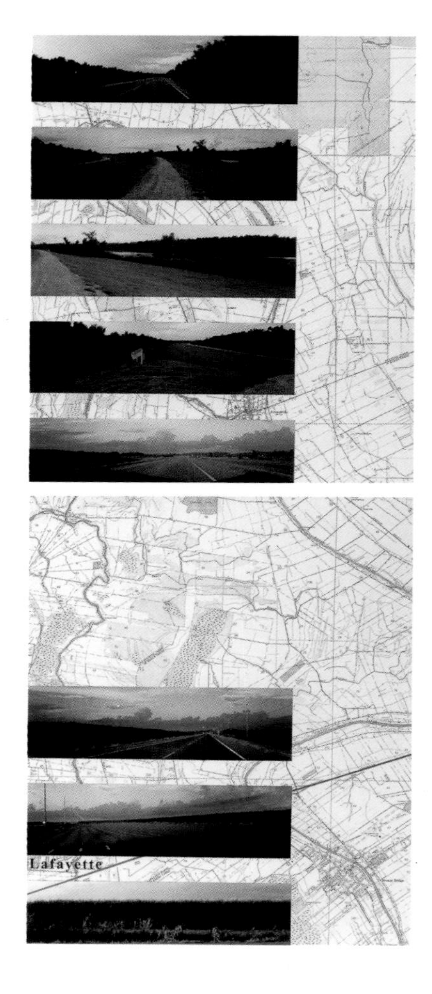

BATON ROUGE TO NEW ORLEANS / RIVER CORRIDOR

BANKS

SITE 3

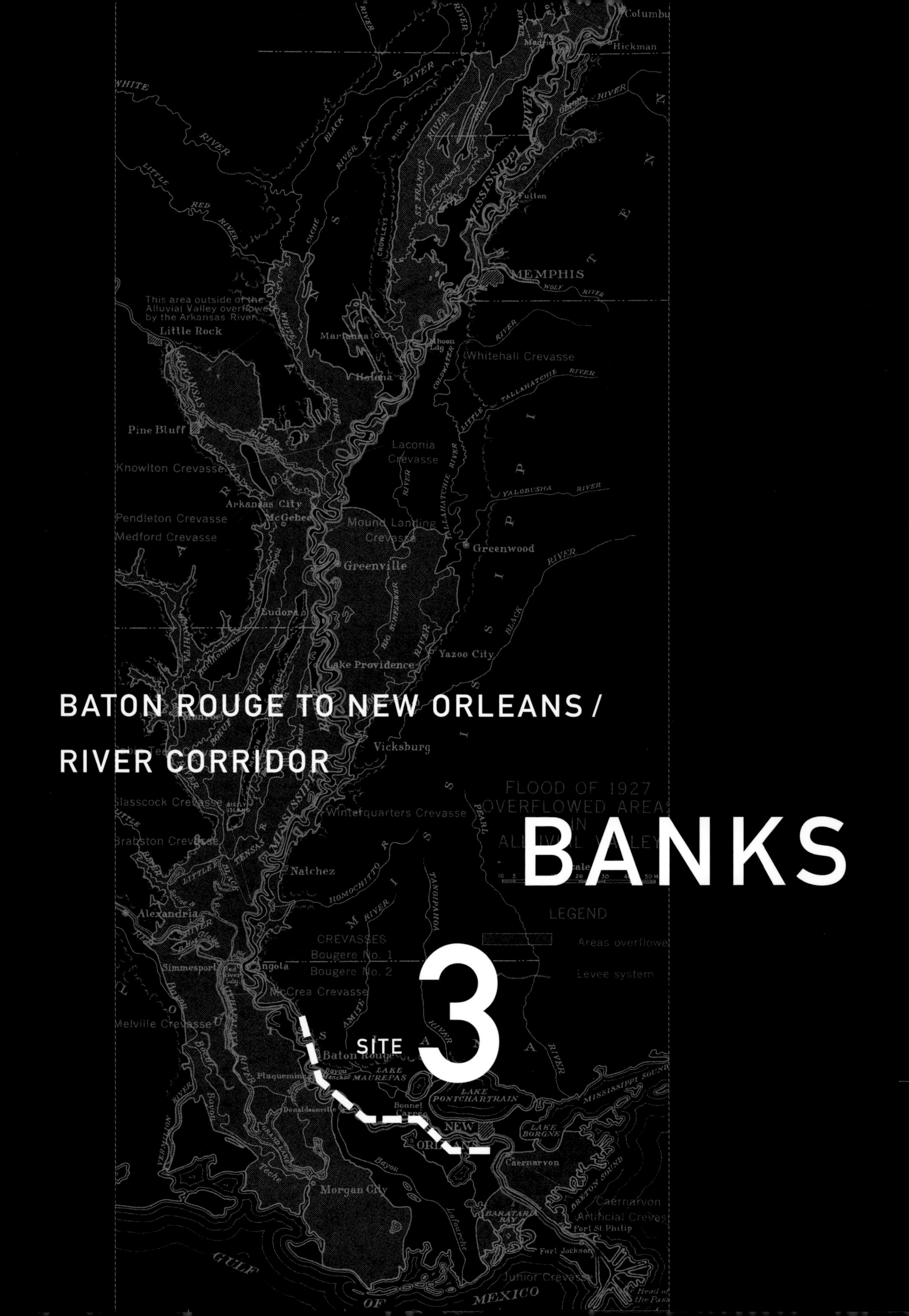

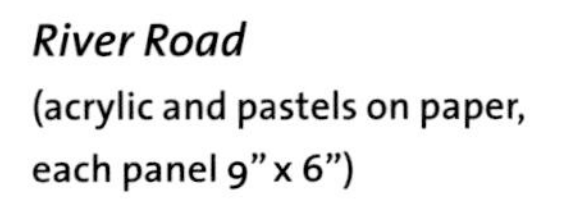

River Road
(acrylic and pastels on paper,
each panel 9" x 6")

CULTIVATING BANKS

From *levée*, meaning raised or elevated, levees are earthen mounds that border almost the entire length of the river, well over 3,500 miles all together in the Lower Mississippi valley. Nowhere, however, are levees more prominent, both materially and culturally, as they are on the "River Road" between Baton Rouge and New Orleans. Here, levees are riverbanks as well as protection walls. They are constructive agents of a linear plantation culture now turned into a booming petrochemical industry corridor, as well as the most cultivated and tended construction in themselves.[1]

When Europeans first chose to settle here in the early 1700s, as with the several Native American cultures that preceded them, natural levees adjacent to the river provided them the only high ground. Natural levees are slim, fertile edges on either side of the river, half a mile to a mile wide. They are built up when the river overflows and deposits its heaviest and coarsest sediment in the immediate vicinity and lays the finer-grained clays and silt on a gradual slope down to backswamps. Over the years the riverside edge of these natural levees had to be further built up to protect settlements against overflow. Initially they were constructed on an individual basis by riparian landowners but quickly gave way to more organized systems of construction as floods intensified with increasing settlement. By 1880 the entire length of the Lower Mississippi, some one thousand miles, was organized into levee districts for the construction and maintenance of levees under the supervision of the Mississippi River Commission. These constructed levees atop natural levees are today more than forty feet high. In a land where high ground is scarce, they run close to the water's edge, giving the river little room to spread laterally. Instead they accommodate the river's variations vertically, often at levels higher than the adjacent land.

From mounds of earth (left) **to engineered walls** (below) [U.S. Army Corps of Engineers]

Drawing recording the changing profile of levees, from D. O. Elliott, ***The Improvement of the Lower Mississippi River for Flood Control and Navigation*****, 1932**

Oak allées and property lines reinforce the arpent measure
[Middle: U.S. Army Corps of Engineers]

The order of the levee is reinforced along its length by landings, loading docks, oceangoing vessels, and a road that follows the river, fronted by small towns. Across its length it is reinforced by the arpent measure. Riparian land grants were based on arpent frontage, with a standard lot depth of forty or eighty arpents that extended properties over a mile into the backswamps. The property lines perpendicular to the levee result in a unique ordering of wedge-shaped properties. Accentuating this property system are allées of oak trees that lead from the river to antebellum plantation homes; drains into the backswamps that made it possible to grow indigo, tobacco, rice, maize, cotton, and, most successfully until the Civil War, sugarcane; and grids of oil tanks and other structures of the petrochemical industry that uses the land of former plantations to produce oil-based synthetic fibers, plastics, fertilizers, pesticides, pharmaceuticals, and pigments.

As much as these cultivations enforce the levee, the levee enforces a cultural order. The River Road is a two-way street. It took a plantation regime equipped with slaves first brought to these parts in 1719 to tame a land "unwonted to European eyes," "fearfully wild," and regularly claimed by both river and sea with embankments that were constantly washed away. Today, it takes their corporate inheritors to do the same, the consequences of their production more diverse but just as controversial, trickling down the embankment to the many disenchanted though now franchised inhabitants who line this "chemical corridor."[2]

The American Ruhr at work around the clock

BATON ROUGE
NEW ORLEANS
oil & gas
tobacco
pharmaceuticals
plastics
petrochemicals
rice
maize
pesticides
sugarcane
pigments

Cultivating Banks
(screen print on paper, 20" x 60")

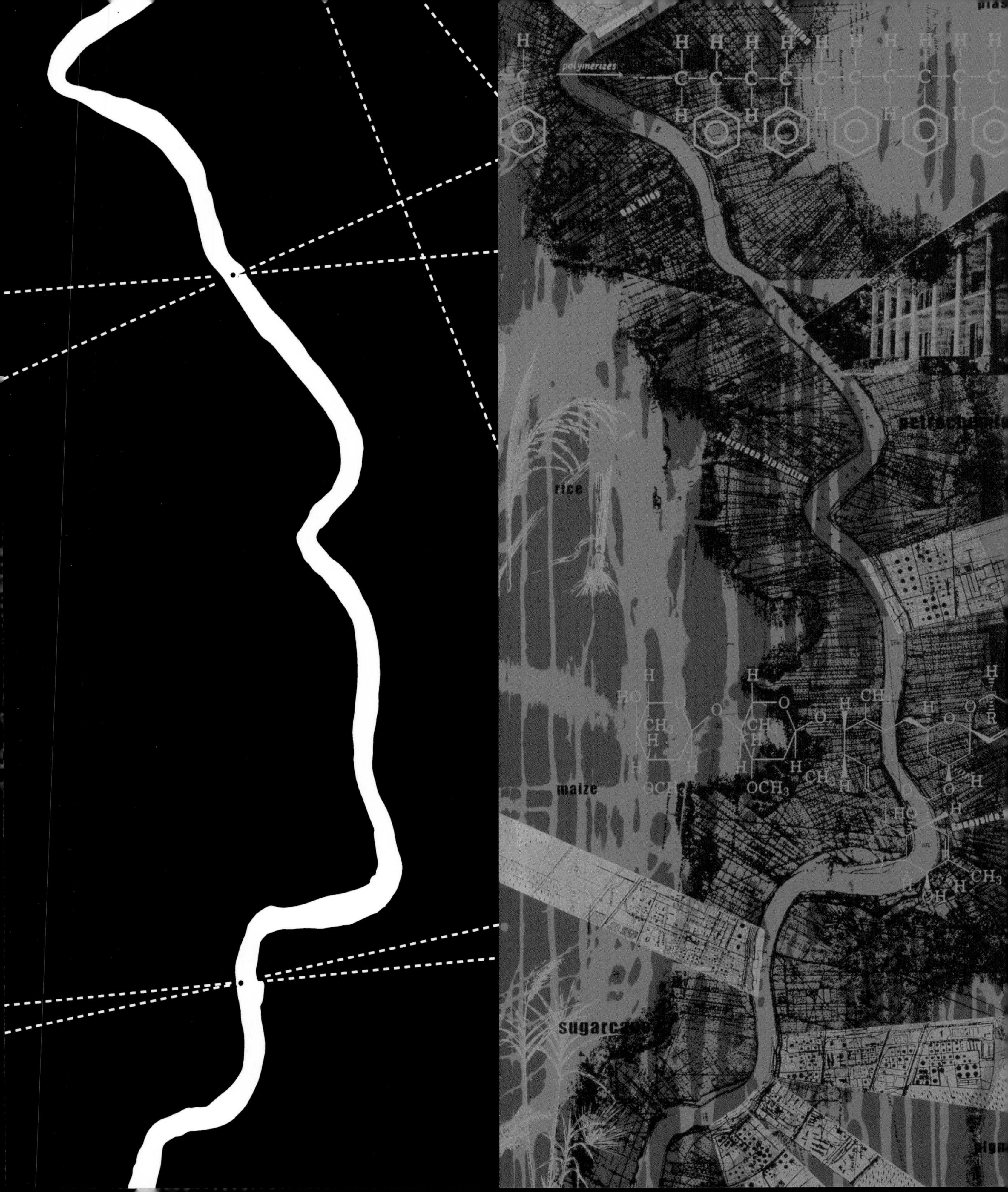
polymerizes
Oak Alley
Evergreen Plantation
rice
maize

CREVASSING LEVEES

Where a crawfish or a muskrat hole can undermine a levee that channels the waters of the third-largest basin in the world, at levels higher than the adjacent land, levees occupy a prominent place in the everyday imagination. Weakened by seepage, the river can come through either in a "boil"—a tiny geyser on the land side of the levee—or in a dam burst that gouges the earth dozens of feet deep. When it is realized that such "crevasses" relieve the pressure on levees up and down the river, levees assume a prominent place in politics as well. Like a border with an enemy country, levees have been patrolled, transgressed, undermined, and monitored over the years with increasing sophistication and centralization.

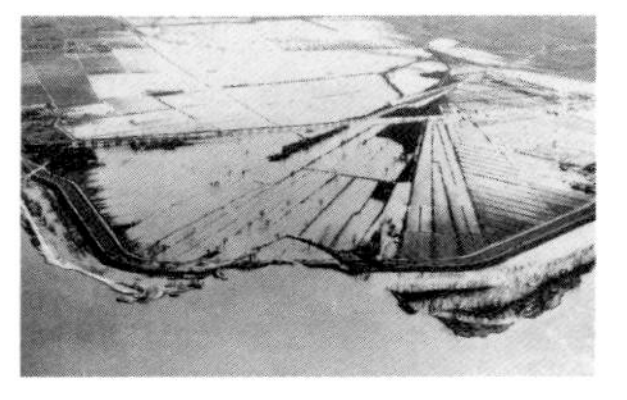

A growing crevasse on the River Road [U.S. Army Corps of Engineers]

Levees have had ample time, funds, private interest, and theory to evolve from the mile-long, yard-high stretch that protected New Orleans when it was a decade old in 1727.[3] Until the Mississippi River Commission decided to employ outlets in addition to levees for flood control after the 1927 flood, the levees-only policy served the local and immediate interest of residents. Velocity, the underpinning of the policy, is particularly important to the Lower Mississippi because for the last 450 miles the riverbed lies below sea level—15 feet below it at Vicksburg, and more than 170 feet below at New Orleans. Given that the river at New Orleans is also approximately 170 feet deep, a large volume of the Lower Mississippi has no reason to flow. It is pushed by the runoff from 1.25 million square miles of continent. The push of a basin results, as John Barry explains, in "a tumbling effect as water spills over itself like an enormous ever-breaking internal wave [that] can attack a riverbank—or a levee—like a buzz saw."[4]

Reinforcing banks with revetments below the water line and concrete pavement above [U.S. Army Corps of Engineers]

Levees are no longer piles of dirt deposited and compacted by slaves under the supervision of riparian landowners, as they were at the turn of the nineteenth century. They are products of a science built and protected by a federal-state partnership under the eye of the Army Corps of Engineers. They are laid on foundations cleared of logs, decayed vegetable matter, and, importantly, taproots thicker than one and a half inches.[5] Their surfaces are shaped to a calculated angle of repose, then covered by well-cropped grass on the outside and paved with concrete on the inside to protect against the wakes thrown by boat traffic. Below the waterline they are lined by revetments that protect the bank from the tumbling effect of the current. Once made of willow mattresses, revetments today are slabs of reinforced concrete four feet long, fourteen inches wide, and three inches thick, tied together with heavy stainless steel wire to form mats. These mats, 25 by 140 feet, are laid by specially designed barges and anchored to the river bank. Their flexibility allows them to take the shape of a bank cleaned and graded by itinerant laborers who live temporarily on dormitory ships docked along the levees.

The need to force a crevasse in these stronger levees to save a city from flood, which had to be done with dynamite at Caernarvon below New Orleans on April 29, 1927—amid promises of compensation, riotous public meetings, and fear of reprisals—has been reduced. Today levees are assisted by outlets. The Caernarvon crevasse, besides blowing away a section of a levee, also blew away the levees-only policy and ushered in outlets as a flood-control measure. These outlets include the Bonnet Carré Spillway about thirty miles upriver from New Orleans, completed in 1932, which is one of four on the Lower Mississippi. When its 350 bays are opened, two million gallons of water pass through the spillway every second into Lake Pontchartrain.

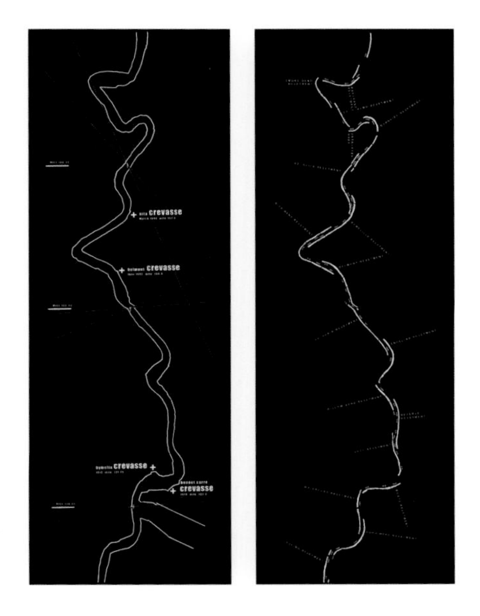

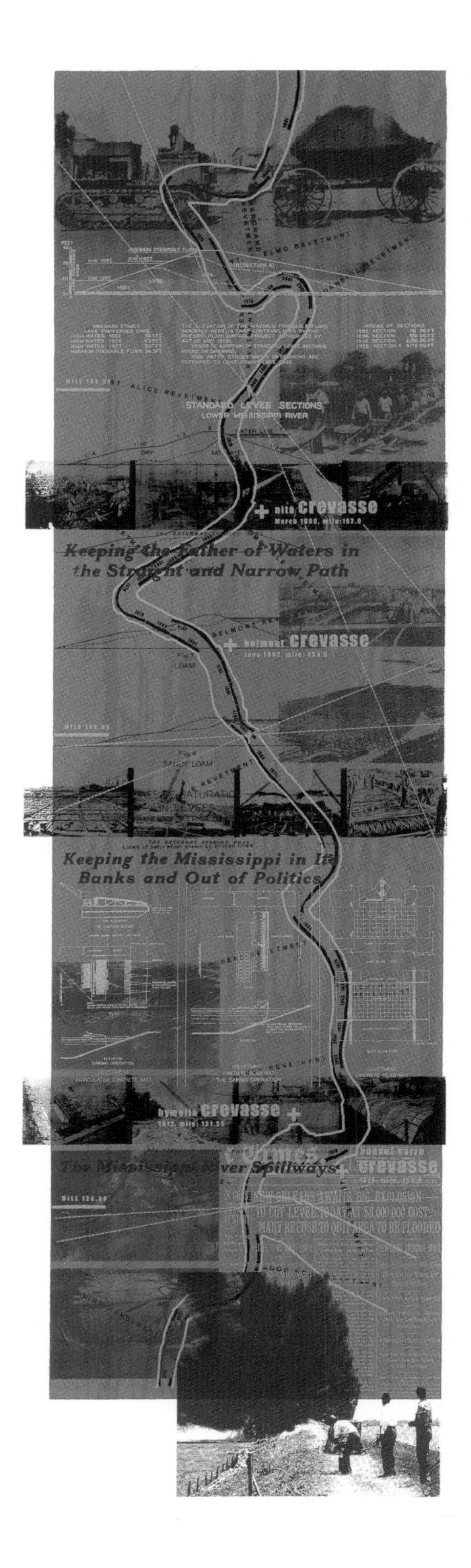

Crevassing Levees
(screen print on paper, 20" x 60")

belmont crevasse
June 1892, mile: 153.0
hymelia crevasse
1912, mile: 131.25
bonnet carre
crevass
1874, mile: 133.0
the Straight and Narrow Path
BELMONT REVETMENT
Fig.3
LOAM
Fig.4
SANDY LOAM
Lines of saturation shown by broken lines
THE SATURDAY EVENING POST
Keeping the Mississippi in Its
Banks and Out of Politics
END ELEVATION
THE SINKING BARGE
PLAN
ELEVATION
SINKING OPERATION
REVETMENT
ARTICULATED CONCRETE MAT
REVETMENT
CONCRETE SLAB MAT
THE SINKING OPERATION
LAP SLAB TYPE
BUTT SLAB TYPE
The Mississippi River Spillways
NEW ORLEANS AWAITS BIG EXPLO
TO CUT LEVEE TODAY AT $2,000,00
MANY REFUSE TO QUIT AREA TO BE
RELIEF FUND INADEQUATE

CHANNELING CROSSINGS

Buoys and bridges frame the shipping channel

The Baton Rouge–New Orleans corridor is a landscape of crossings—crossings along the flow of the Mississippi as well as across it. Crossings along the flow of the river are tracked by ships as they follow a trajectory that seems to exaggerate the river's bends. They are following the river's deepest waters, waters that cross sides on the reaches—straight stretches between bends—as the eroding and depositing banks of a twisting and turning river switch sides. The point where this switch occurs—where the cross section of the river is roughly symmetrical—is called a crossing. Making a crossing, ships ride the strongest waters of the river, waters that scour the outer banks of bends. These waters cannot easily change the course of the river as they once regularly did, for banks are well protected with concrete revetments. Instead their energy is concentrated on moving sediment and maintaining a channel for ships to and from ports around the world. It is a path marked by range lights, buoys, shore lights, depth finders, and regular radio updates of river stages by Coast Guard ships patrolling these waters, a landscape largely unseen by those not trained to see its order.

Ships and barges share the same channel

It is the responsibility of the Corps of Engineers to maintain a navigable channel five hundred feet wide and thirty-five feet deep between Baton Rouge and New Orleans for barges and deep-ocean vessels, and a channel three hundred feet wide and nine feet deep above Baton Rouge. Ingeniously, the Corps has put the river to work dredging its own channel that the engineers have only to watch and at times assist. Assistance is in the form of jetties, walls of piled stone

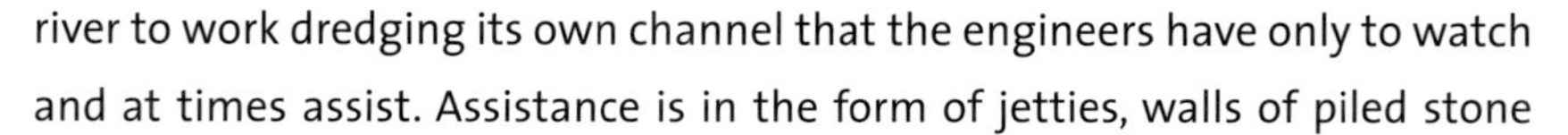

extending the right distance into the current from banks and islands and built to a height just above the low-water line. They urge the stream to maintain navigable depths while consolidating land on their upstream side. With soundings the Corps keeps a constant vigil on the hidden topography scoured by these working waters.

Early ferry crossings powered by mules
[Captain Riley Powell]

If ships amplify the bends of the River Road when they plot its deepest waters, then the three bridges between Baton Rouge and New Orleans heighten the gentle slope of the land that it divides when they begin their rise out of the fields and backswamps over a mile away to cross above river traffic. The intermittent crossings of ferries once powered by mules from landings that were hubs of activity have decreased. Ferries connect people of a parish (the Louisiana equivalent of a county) that the Mississippi, interestingly, does not border but divides. It was evidently easier for settlers to paddle across the Big Muddy than up and down its waters.[6]

The least visible crossings are the pipelines of oil and gas that feed the heavy industries lining the river today. They cross under the riverbed, the only sign of their presence being a notice that reads, "Do not anchor or dredge." But the danger these pipelines face is perhaps less from anchors and dredges than from the scouring and deepening waters of the Mississippi making their crossings, pushing past New Orleans at 304 billion gallons per day.

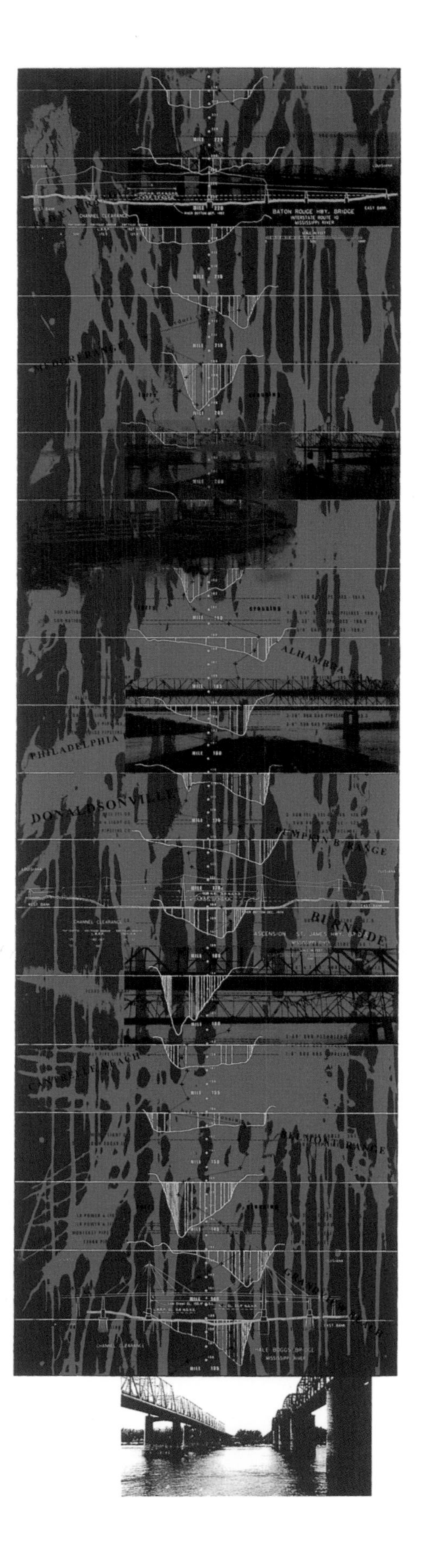

Channeling Crossings
(screen print on paper, 20" x 60")

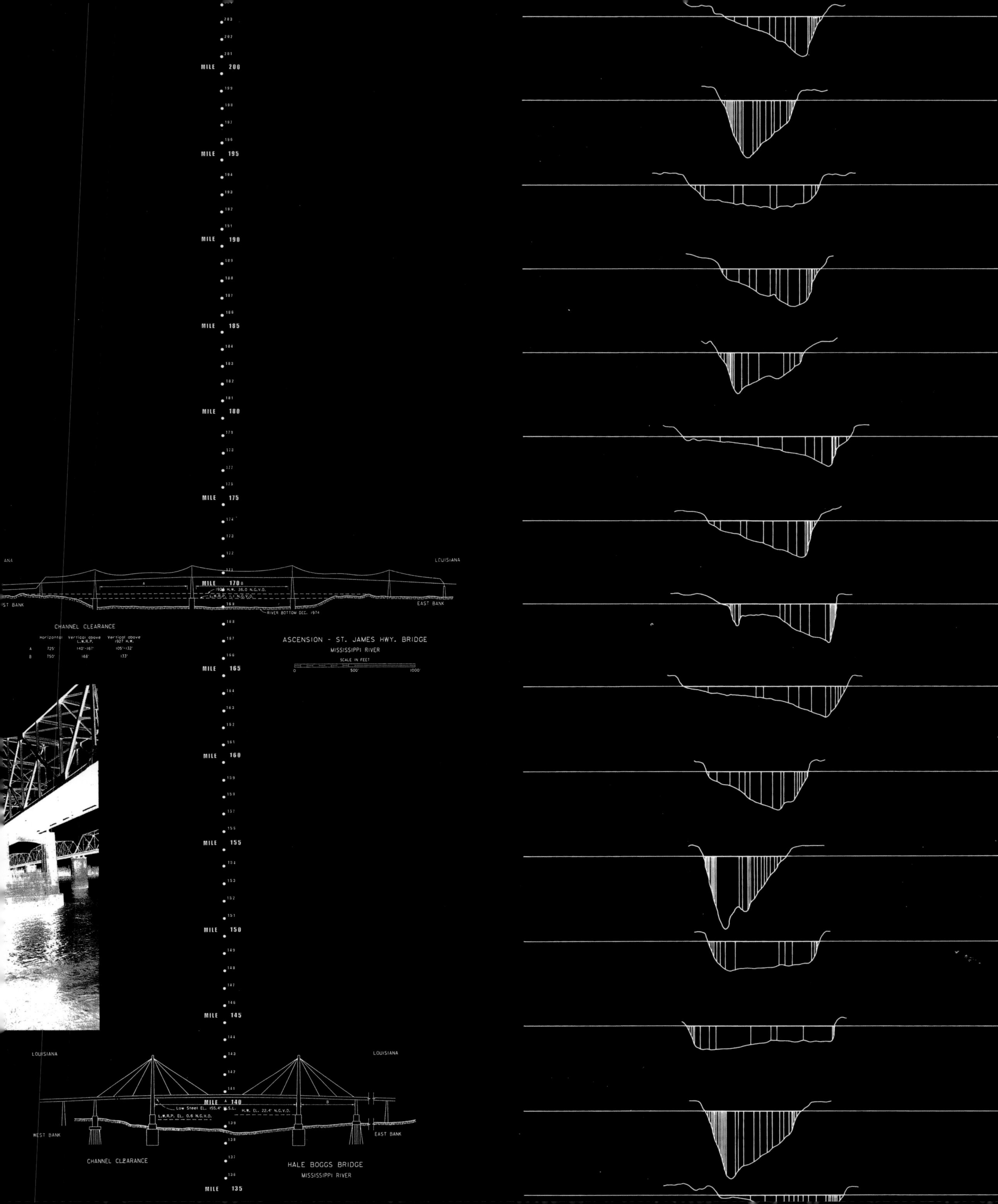
MILE 200
MILE 195
MILE 190
MILE 185
MILE 180
MILE 175
LOUISIANA
MILE 170
1927 H.W. 36.0 N.G.V.D.
L.W.R.P. 1.1 N.G.V.D.
EAST BANK
RIVER BOTTOM DEC. 1974
CHANNEL CLEARANCE
Horizontal | Vertical above L.W.R.P. | Vertical above 1927 H.W.
A 725' 140'-167' 105'-132'
B 750' 168' 133'
ASCENSION - ST. JAMES HWY. BRIDGE
MISSISSIPPI RIVER
SCALE IN FEET
0 500' 1000'
MILE 165
MILE 160
MILE 155
MILE 150
MILE 145
LOUISIANA
LOUISIANA
MILE 140
Low Steel EL. 155.4' M.S.L.
H.W. EL. 22.4' N.G.V.D.
L.W.R.P. EL. 0.6 N.G.V.D.
WEST BANK
EAST BANK
CHANNEL CLEARANCE
HALE BOGGS BRIDGE
MISSISSIPPI RIVER
MILE 135

FLOATING MATTERS

The many assemblages of barges that ply the Mississippi today have come a long way from the "mighty rafts" that Mark Twain saw gliding by during his boyhood in Hannibal. Those rafts were "an acre or so of white, sweet-smelling boards in each raft, a crew of two dozen men or more, three or four wigwams scattered about the raft's vast level space for storm quarters."[7] Tows, as today's assemblages of barges are called, are as large as six acres. They are landscapes held together by steel cables an inch thick that have been known to snap like rubber bands in rough situations. Some tows are hilly with twenty- to thirty-foot mounds of coal, half of which are contained below the waterline. Others are flat, covered containers of corn, steel pipes, gypsum, and a range of other manufactured and raw materials being shipped to and from ports in the Mississippi basin. Most are combinations. A tow is managed by half a dozen workers, that is, one person to an acre, making it more than twenty-four times as efficient as the mighty rafts of Twain's time. Considering that the six people on a tow manage a load that would fill three thousand trucks, they make the waterways many times more efficient than roadways as they orchestrate the addition and subtraction of barges and maintain the varying tensions of their connectors.

Tows in tension carry vast amounts of material on the Mississippi

The changing configuration and size of a tow reflects the complexities of a vast navigational network. Already in the 1930s the network was connecting places as far-flung as Pittsburgh, Pennsylvania, and St. Croix, Minnesota, with the Gulf trade. Tows are pushed up the Missouri, Mississippi, Ohio, and Illinois rivers, gradually narrowing down from the mile-wide Lower Mississippi, and climbing with the help of twenty-seven locks and dams that maintain depths of nine to twelve feet. The hinterland is extended by the Intracoastal Waterway, a protected canal along the Gulf coast built by the Corps of Engineers in the first half of the 1900s. Referred to by the Corps as the "crossing of the T of trade," the waterway connects New Orleans with Houston and Brownsville in Texas and Pensacola in Florida.

Only a short time before Twain, the Mississippi was a one-way road, until, that is, 1812, when the first steamboat arrived in New Orleans. Until the 1850s flatboats—or "floating boxes"—came down the Mississippi with hides, furs, farm products, and the wood of which they were made.[8] At New Orleans these forerunners of barges were broken up and sold for lumber. The crew stayed on or walked back up the river. In these early days of navigation, men competed for space on the river with more anarchic woody matter washed off the banks by a meandering Mississippi, mainly snags. Snags referred to either "planters," which were trees fixed in the bed of the river, or "sawyers," trees that rose and fell on the surface in a position to spear a boat coming upstream. They were a constant threat to boatmen and were particularly troublesome on tributaries for the rafts they sometimes initiated. These rafts did not move. They merely grew, collecting masses of floating debris and driftwood that could stretch for many miles, like the raft on the Red River forty miles long that was eventually broken up by Captain Henry Shreve, the inventor of the snagboat that could lift huge trees out of the water.

With all the bank protection and vigilance on the Mississippi, today's barges made of steel are hardly threatened by the danger of snags. They glide confidently past oceangoing vessels less than half their length.

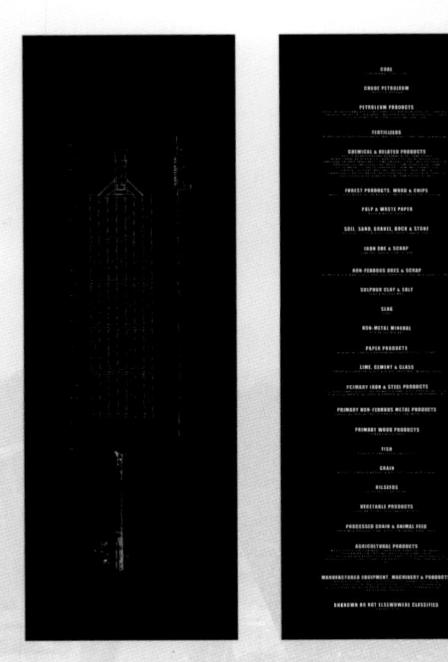

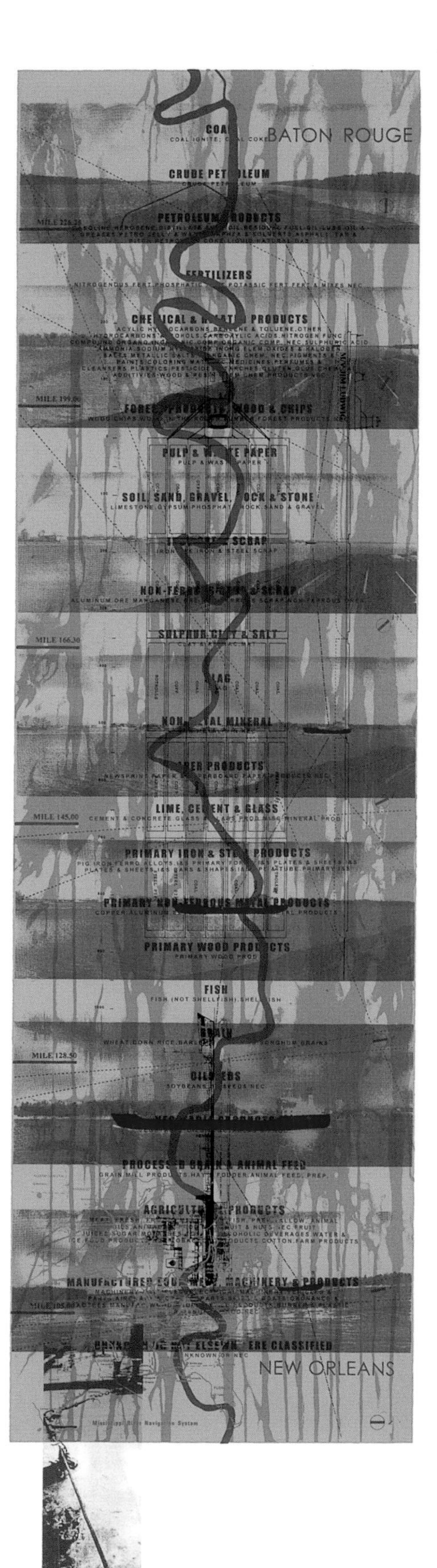

Floating Matters
(screen print on paper, 20" x 60")

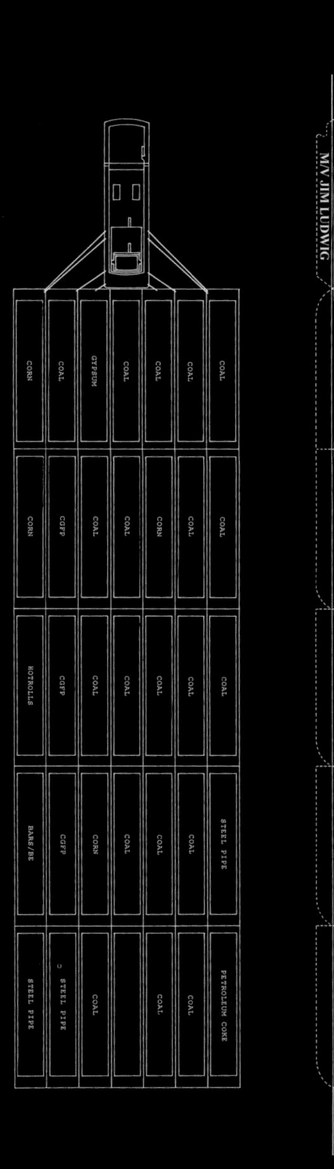

COAL
COAL IGNITE; COAL COKE

CRUDE PETROLEUM
CRUDE PETROLEUM

PETROLEUM PRODUCTS
GASOLINE.KEROSENE.DISTILLATE FUEL OIL.RESIDUAL FUEL OIL.LUBE OIL & GREASES.PETRO JELLY & WAXES.NAPHTA & SOLVENTS.ASPHALT, TAR & PITCH.PETROLEUM COKE.LIQUID NATURAL GAS

FERTILIZERS
NITROGENOUS FERT.PHOSPHATIC FERT.POTASSIC FERT.FERT & MIXES NEC

CHEMICAL & RELATED PRODUCTS
ACYLIC HYDROCARBONS.BENZENE & TOLUENE.OTHER HYDROCARBONS.ALCOHOLS.CARBOXYLIC ACIDS.NITROGEN FUNC. COMPOUND.ORGANO INORGANIC COMP.ORGANIC COMP. NEC.SULPHURIC ACID AMMONIA.SODIUM HYDROXIDE.INORG.ELEM.OXIDES & HALOGEN SALTS.METALLIC SALTS.INORGANIC CHEM. NEC.PIGMENTS & PAINTS.COLORING MAT. NEC.MEDICINES.PERFUMES & CLEANSERS.PLASTICS.PESTICIDES.STARCHES.GLUTEN.GLUE.CHEMICAL ADDITIVIES.WOOD & RESIN CHEM.CHEM.PRODUCTS NEC

FOREST PRODUCTS, WOOD & CHIPS
WOOD CHIPS.WOOD IN THE ROUGH.LUMBER.FOREST PRODUCTS NEC

PULP & WASTE PAPER
PULP & WASTE PAPER

SOIL, SAND, GRAVEL, ROCK & STONE
LIMESTONE.GYPSUM.PHOSPHATE ROCK.SAND & GRAVEL

IRON ORE & SCRAP
IRON ORE.IRON & STEEL SCRAP

NON-FERROUS ORES & SCRAP
ALUMINUM ORE.MANGANESE ORE.NON-FERROUS SCRAP.NON-FERROUS ORES

SULPHUR CLAY & SALT
CLAY & REFRAC.MAT.

SLAG
SLAG

NON-METAL MINERAL
NON-METAL.MIN.NEC

PAPER PRODUCTS
NEWSPRINT.PAPER & PAPERBOARD.PAPER PRODUCTS NEC

LIME, CEMENT & GLASS
CEMENT & CONCRETE.GLASS & GLASS PROD.MISC.MINERAL PROD.

PRIMARY IRON & STEEL PRODUCTS
PIG IRON.FERRO ALLOYS.I&S PRIMARY FORMS.I&S PLATES & SHEETS.I&S PLATES & SHEETS.I&S BARS & SHAPES.I&SPIPE & TUBE.PRIMARY I&S

PRIMARY NON-FERROUS METAL PRODUCTS
COPPER.ALUMINUM.SMELTED PROD NEC.TAB.METAL PRODUCTS

PRIMARY WOOD PRODUCTS
PRIMARY WOOD PROD.

FISH
FISH (NOT SHELLFISH).SHELLFISH

GRAIN
WHEAT.CORN.RICE.BARLEY & RYE.OATS.SORGHUM GRAINS

OILSEEDS
SOYBEANS.OILSEEDS NEC

VEGETABLE PRODUCTS
VEGETABLE OILS.VEGETABLES & PROD.

PROCESSED GRAIN & ANIMAL FEED
GRAIN MILL PRODUCTS.HAY & FODDER.ANIMAL FEED, PREP.

AGRICULTURAL PRODUCTS
MEAT, FRESH, FROZEN.MEAT, PREP.FISH, PREP.TALLOW, ANIMAL OILS.ANIMALS & PROD NEC.FRUIT & NUTS NEC.FRUIT JUICES.SUGAR.MOLASSES.COFFEE.ALCOHOLIC BEVERAGES.WATER & ICE.FOOD PRODUCTS NEC.TOBACCO & PRODUCTS.COTTON.FARM PRODUCTS NEC

MANUFACTURED EQUIPMENT, MACHINERY & PRODUCTS
MACHINERY (NOT ELEC).ELECTRICAL MACHINERY.VEHICLES & PARTS.AIRCRAFT & CRAFT & PARTS.SHIPS & BOATS.ORDNANCE & ACCESS.MANUFAC.WOOD PROD.TEXTILE PRODUCTS.RUBBER & PLASTIC PR.MANUFAC.PROD.NEC

UNKNOWN OR NOT ELSEWHWERE CLASSIFIED
UNKNOWN OR NEC

RANGING SIGHTS

The pilot, fifty feet above and half a mile behind the front end of the tow

It is difficult to imagine memorizing the course of the Mississippi. But this is a requirement for river pilots. Mark Twain was told that to be a pilot on the Mississippi you had to know the shape of the river "with such absolute certainty that you can always steer by the shape that's *in your head,* and never mind the one that's before your eyes." The eye is liable to be beguiled by light, shadows, and shapes that appear under the many conditions of starlight, moonlight, pitch-darkness, mist, and so on, unless one knows better. The shape Twain refers to is more than just a plan. It included heights of banks, depths of water, marks, bars, towns, islands, bends, reaches. A pilot's memory, particularly with a "villainous river" whose "alluvial banks cave and change constantly, whose snags are always hunting up new quarters, whose sandbars are never at rest, whose channels are forever dodging and shirking," is the basis for reading a shifting landscape. It is to see the sediment deposited on a shelf as not there before, to conclude from differences between now and then. But it is also the basis for inhabiting that landscape. Memory allows a pilot to make changes in the shape of the river in his head without having to rely on his eyes, thus prepared for its new form on a pitch-dark night or when it is "shoreless," as in flood. "With what scorn a pilot was looked upon, in the old times, if he ever ventured to deal in that feeble phrase 'I think,' instead of the vigorous one, 'I know!' "[9]

Radar, with a range of five to ten miles day and night

Memory still serves pilots though they enjoy the assistance of onboard radar images, updated maps, and radio information. Technology not only assists in reinforcing the shape of the river that's in the pilot's head, it also presents a range that is in proportion to today's

vessels. To the pilot in the wheelhouse high enough to see the two corners at the head of his tow a quarter mile away, radar seems crucial. The radius of the image that he sees on his radar screen is about five to ten miles long. Unlike the driver of a vehicle who moves in the landscape map in hand, the radar moves the landscape. Landscapes enter and exit the screen or rotate around the pilot whose position and alignment is fixed in the blip at its center that faces the top of the screen.

Night and day are the same on this radar screen, and xenon beams mounted on the tow can light up precise spots of land many miles away. Shore lights, channel buoys, and channel reports make navigation safer than it was for Twain. But as modern instruments and communication make the maneuvering of boats easier, the demands of global economics for greater and greater tonnage per tow continue to push the limits of the captain's sight.

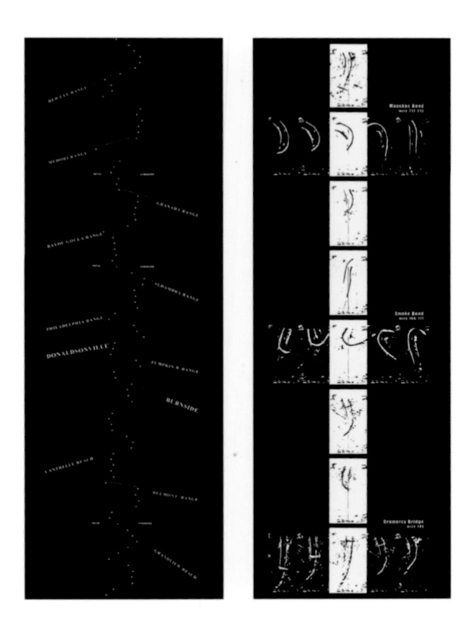

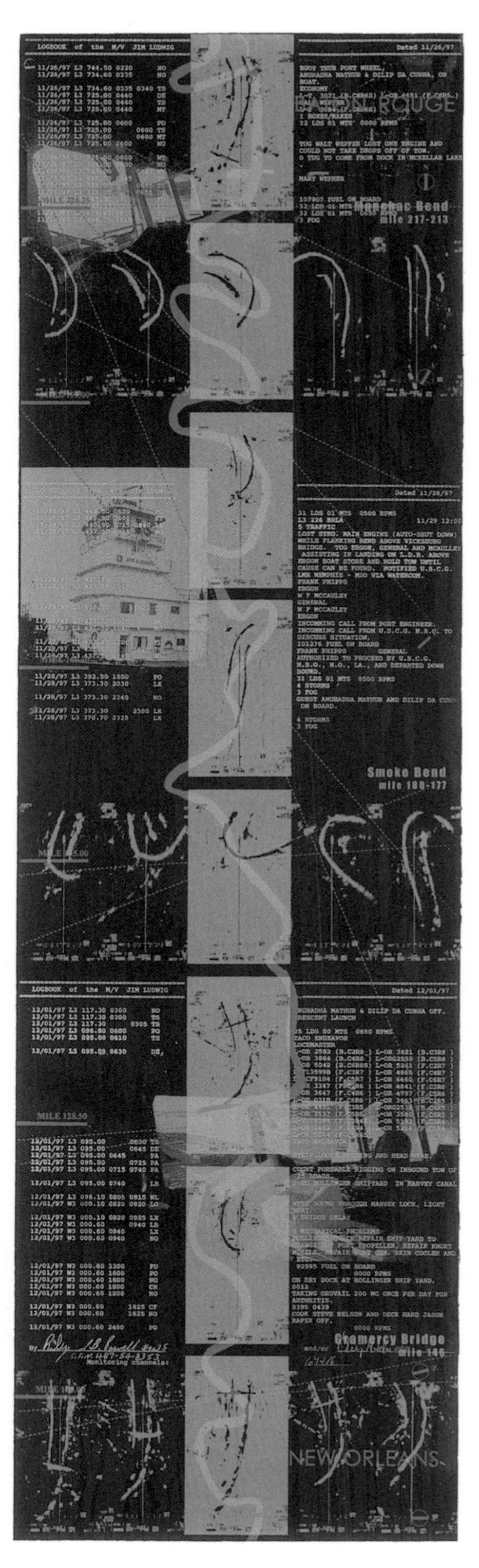

Ranging Sights
(screen print on paper, 20" x 60")

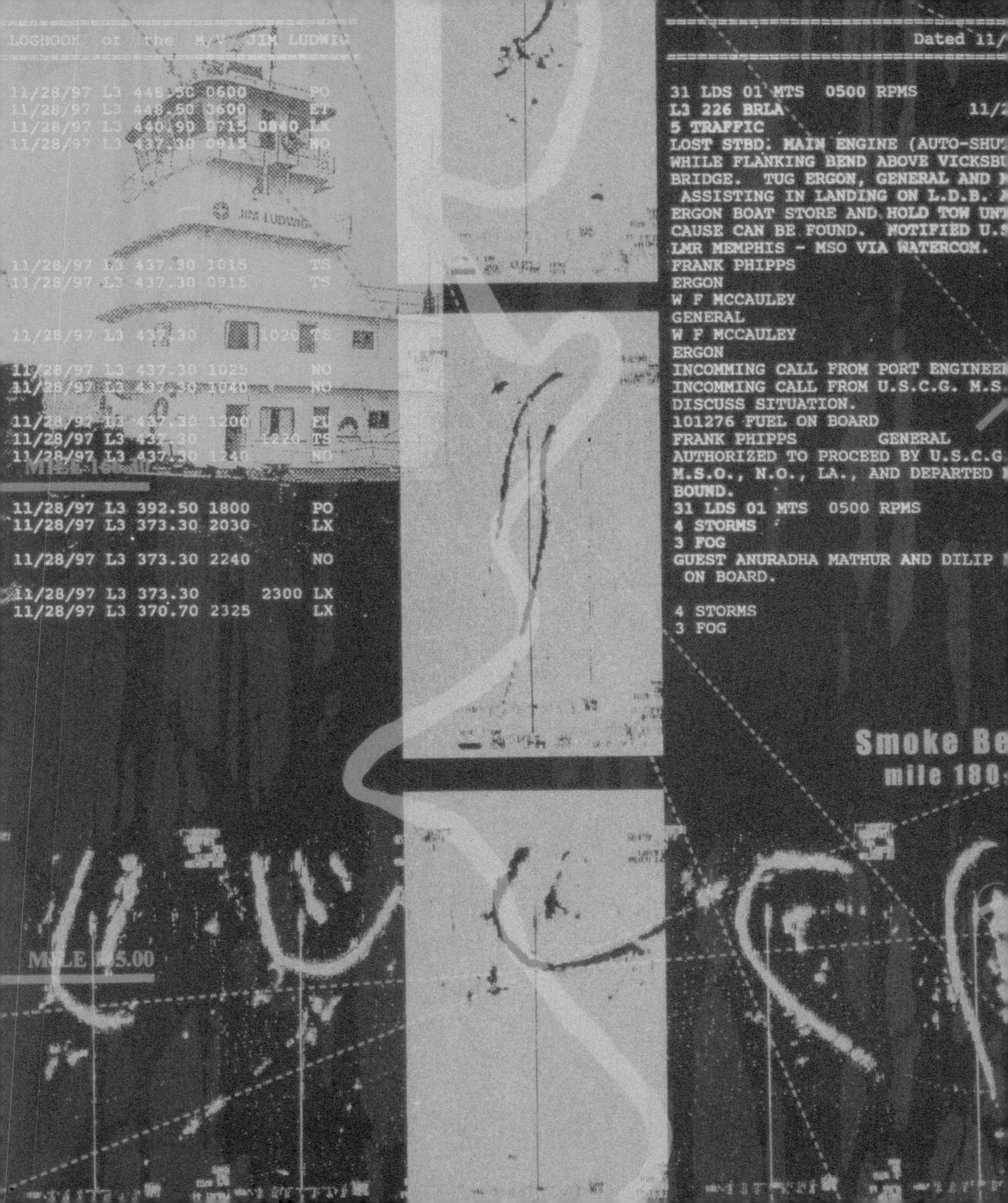

LOGBOOK of the M/V JIM LUDWIG

11/28/97 L3 448.50 0600 PO
11/28/97 L3 448.50 0600 ET
11/28/97 L3 440.90 0715 0840 LX
11/28/97 L3 437.30 0915 NO

JIM LUDWIG

11/28/97 L3 437.30 1015 TS
11/28/97 L3 437.30 0915 TS

11/28/97 L3 437.30 1020 TS

11/28/97 L3 437.30 1025 NO
11/28/97 L3 437.30 1040 NO

11/28/97 L3 437.30 1200 EU
11/28/97 L3 437.30 1220 TS
11/28/97 L3 437.30 1240 NO

11/28/97 L3 392.50 1800 PO
11/28/97 L3 373.30 2030 LX

11/28/97 L3 373.30 2240 NO

11/28/97 L3 373.30 2300 LX
11/28/97 L3 370.70 2325 LX

MILE 365.00

Dated 11/

31 LDS 01 MTS 0500 RPMS
L3 226 BRLA 11/2
5 TRAFFIC
LOST STBD. MAIN ENGINE (AUTO-SHU
WHILE FLANKING BEND ABOVE VICKSBU
BRIDGE. TUG ERGON, GENERAL AND
ASSISTING IN LANDING ON L.D.B.
ERGON BOAT STORE AND HOLD TOW UNT
CAUSE CAN BE FOUND. NOTIFIED U.S
LMR MEMPHIS - MSO VIA WATERCOM.
FRANK PHIPPS
ERGON
W F MCCAULEY
GENERAL
W F MCCAULEY
ERGON
INCOMMING CALL FROM PORT ENGINEER
INCOMMING CALL FROM U.S.C.G. M.S.
DISCUSS SITUATION.
101276 FUEL ON BOARD
FRANK PHIPPS GENERAL
AUTHORIZED TO PROCEED BY U.S.C.G
M.S.O., N.O., LA., AND DEPARTED
BOUND.
31 LDS 01 MTS 0500 RPMS
4 STORMS
3 FOG
GUEST ANURADHA MATHUR AND DILIP
ON BOARD.

4 STORMS
3 FOG

Smoke Be
mile 180

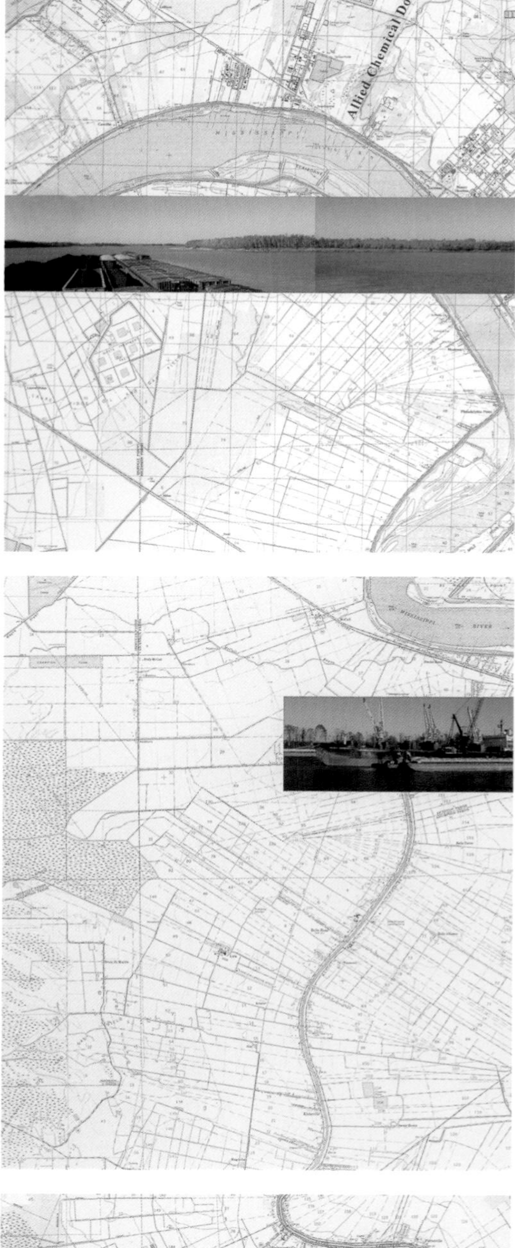

Between Earthen Shores
(color photo prints
on USGS maps,
each panel 15" x 17")

Going down the River Road (the stretch of the river from Baton Rouge to New Orleans) on the tugboat MV *Jim Ludwig* pushing a tow of forty barges—nearly six football fields in area—the landscape is strangely dwarfed until you pass under a bridge. Then the tow seems to grow to its full quarter-mile length, gliding by like a huge spaceship. As the pilot skillfully maneuvers between buoys and lights and through bends and reaches, fog, rain, and near collisions, the horizon unfolds at eight knots between levees about three thousand feet apart. The channeling effect of the levees is heightened by rhythmic landings, docks, and oceangoing vessels that harbor along the length of the River Road. At night these structures make a wall of lights. This segment of the river, also known as the American Ruhr, is a booming round-the-clock petro chemical and transportation industry.

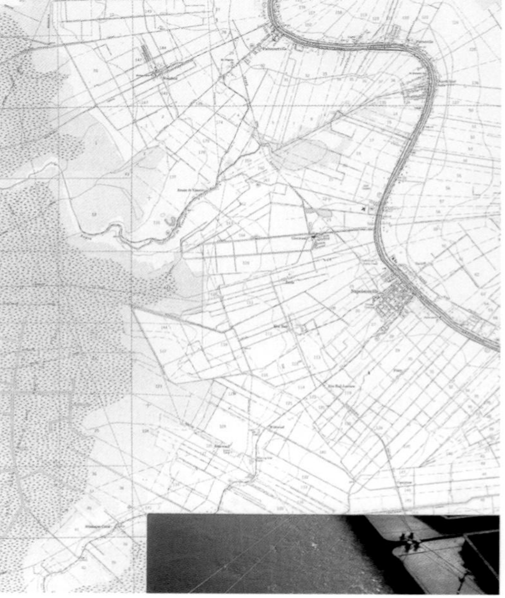

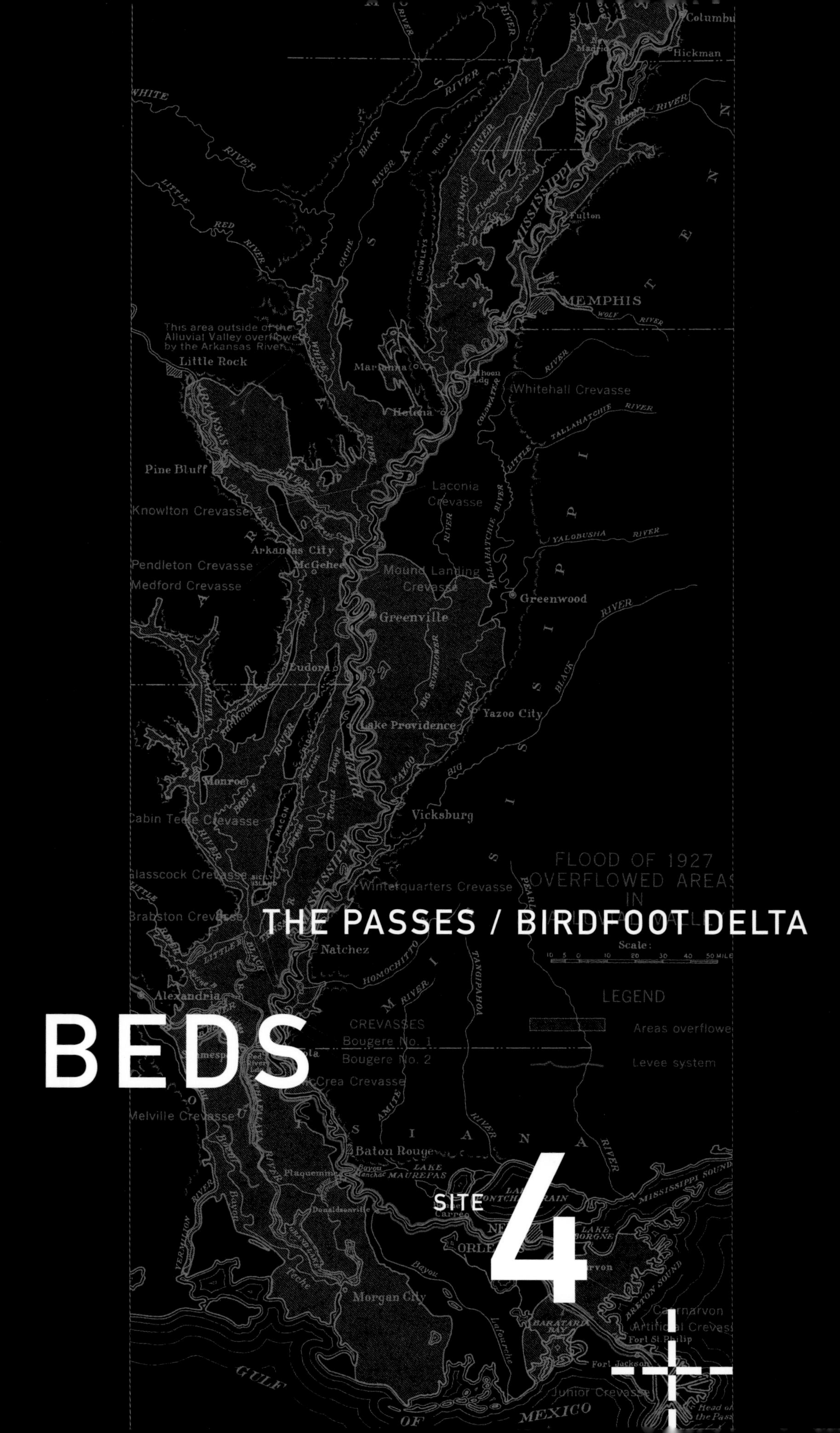

BEDS

SITE 4

THE PASSES / BIRDFOOT DELTA

Land's End
(acrylic and pastels on paper,
each panel 9" x 12")

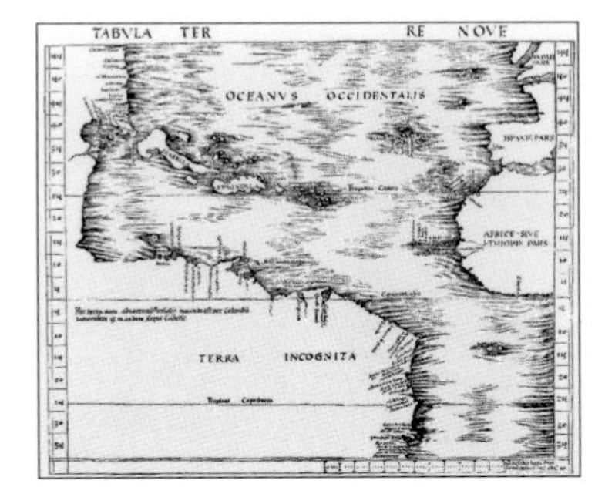

Martin Waldseemüller's map *Tabula Terre Nove* (1513 edition)
[The Historic New Orleans Collection, accession no. 1976.143]

PASSING DEPTH

By the same route that the mud of the continent washes into the Gulf, seafaring vessels are able to enter the heart of the American continent. But it took a while before anyone could brave the passage upriver and as long a while before a way was found to sustain that passage. Columbus, seeing the Mississippi Delta on his fourth voyage, called it the River of Palms. It is the "birdfoot" that appeared in Martin Waldseemüller's map *Tabula Terre Nove*, first published in 1507, with the name that Waldseemüller gave to the New World, America, after Amerigo Vespucci. The delta was noticed by other passersby, too. Pánfilo de Narváez in 1528 made note of "a very great river" where they "take sweet water from the sea." The Mississippi evidently pushes out into the Gulf and into the path of explorers. But it was not until 1699 that the French explorer Iberville entered this far-reaching mouth through which at least three expeditions had emerged but not entered. And it was not until James Eads, a civil engineer, came on the scene in the 1870s that a sustained passage inland though the passes could become a reality.[1]

A hundred miles below New Orleans the Mississippi begins to slow down as it approaches a territory marked by the Corps of Engineers as "Mile 0." Here, the gradually narrowing river divides into three main channels—Southwest Pass, Pass à Loutre, and South Pass. Soundings in the 1870s revealed the depth of the bed at the three passes to be 13, 11, and 8 feet, respectively. For a river bed that is more than 150 feet deep at New Orleans, this shallow clearance, along with the unpredictable formation of sand bars and "mud lumps," presented a major impediment for vessels and consequently for the development of American trade. Mud lumps are "submerged mounds or low, irregular, ephemeral, emergent islands of plastic clay." They are thought to be pushed up by gaseous forces or pressures of advancing bars. They rise sometimes 10 to 15 feet above the water, and vary in size from a few feet in diameter to an area of thirty acres. Despite considerable effort and expense, dredging, common in other parts of the Mississippi valley, failed at the passes. It failed to loosen mud lumps and maintain a dependable passage to the sea. A number of engineers and politicians argued that a navigation canal to the Gulf with elaborate locks and dams, despite its great expense, was justified, until Eads put forth his alternative of jetties.[2]

SS *Manhattan* at the end of the jetties
[U.S. Army Corps of Engineers]

Eads Jetties work on the simple principle of letting the river maintain its own channel. Parallel wall-like structures in the river mouth made of layers of willow mattresses, weighed down by stone and a concrete cap, pinch the current. The resulting acceleration in the water scours the bed to navigable depths. It took great political and financial ingenuity on the part of Eads, a self-taught engineer, to open the "road to the sea," a task that had defeated many engineers before him. Although Eads constructed them initially at his own expense, he was paid a sum of money by the U.S. government after 28 feet of depth was conclusively achieved at South Pass, and further sums for every two feet thereafter. The significance of his achievement is evident in the numbers. The tonnage of goods shipped from St. Louis to Europe increased from 6,857 in 1875, the year he began his work, to 453,681, when he finished, catapulting New Orleans from ninth to second place among United States ports. Eads Jetties extended the reach of the Mississippi.[3]

Restoration of jetties in the Southwest Pass
[U.S. Army Corps of Engineers]

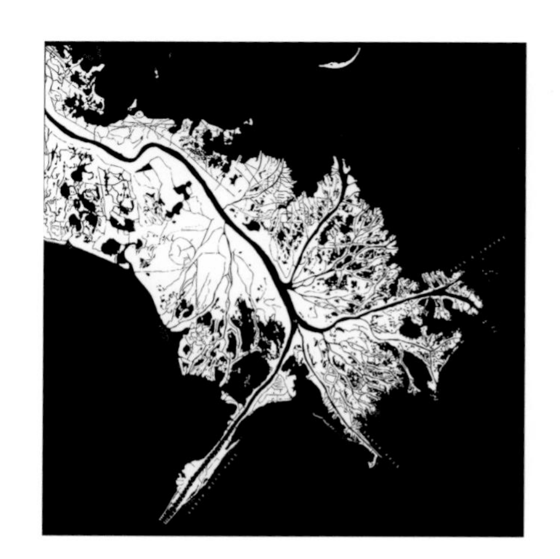

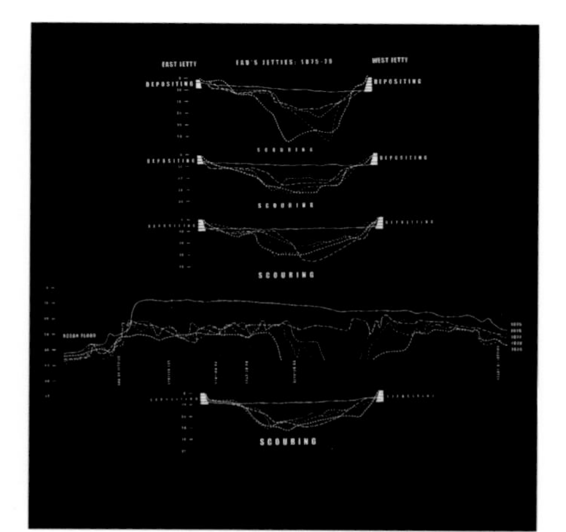

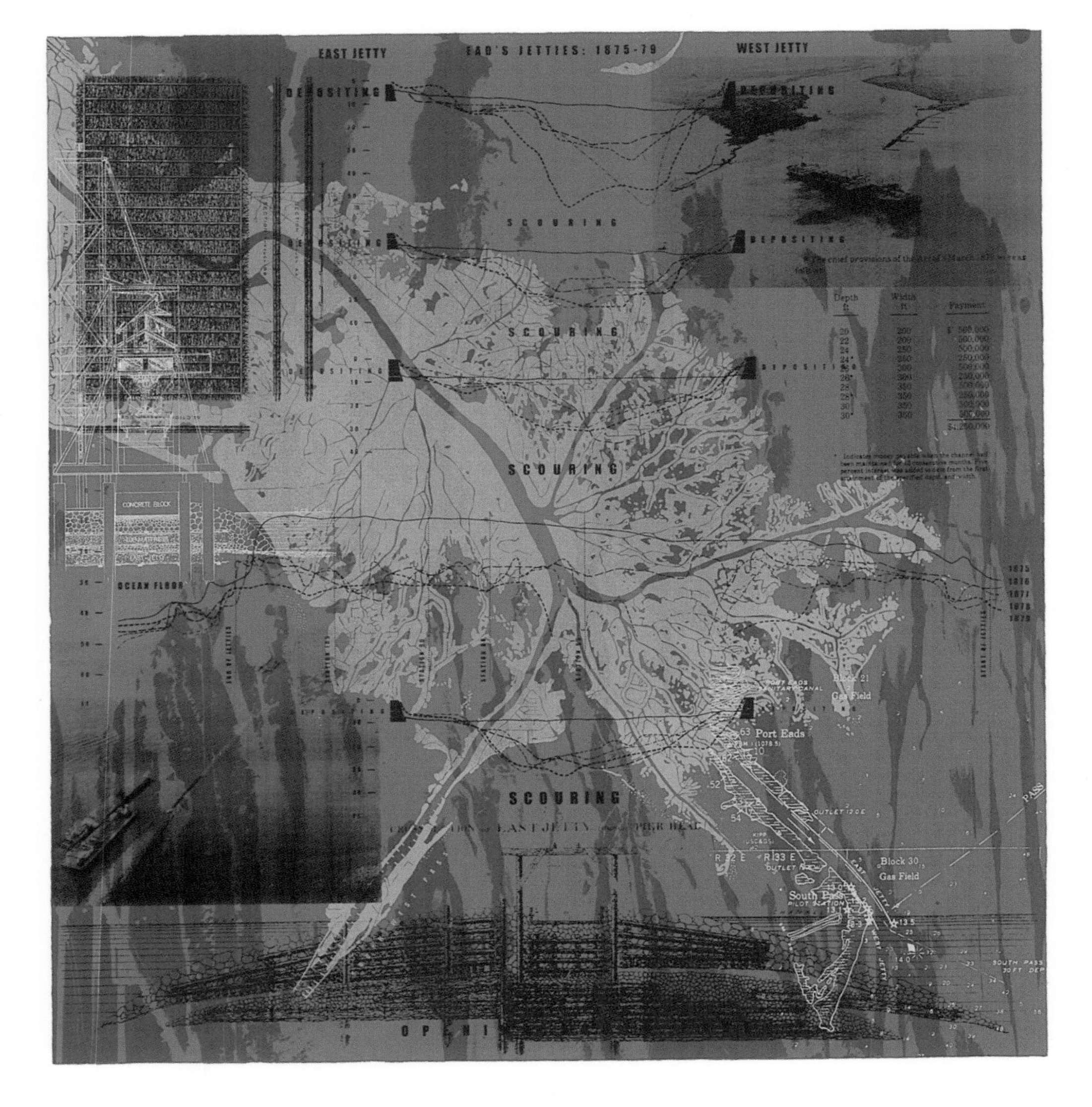

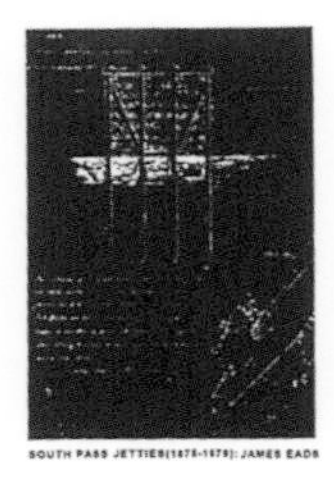

Passing Depth
(screen print on paper, 44" x 30")

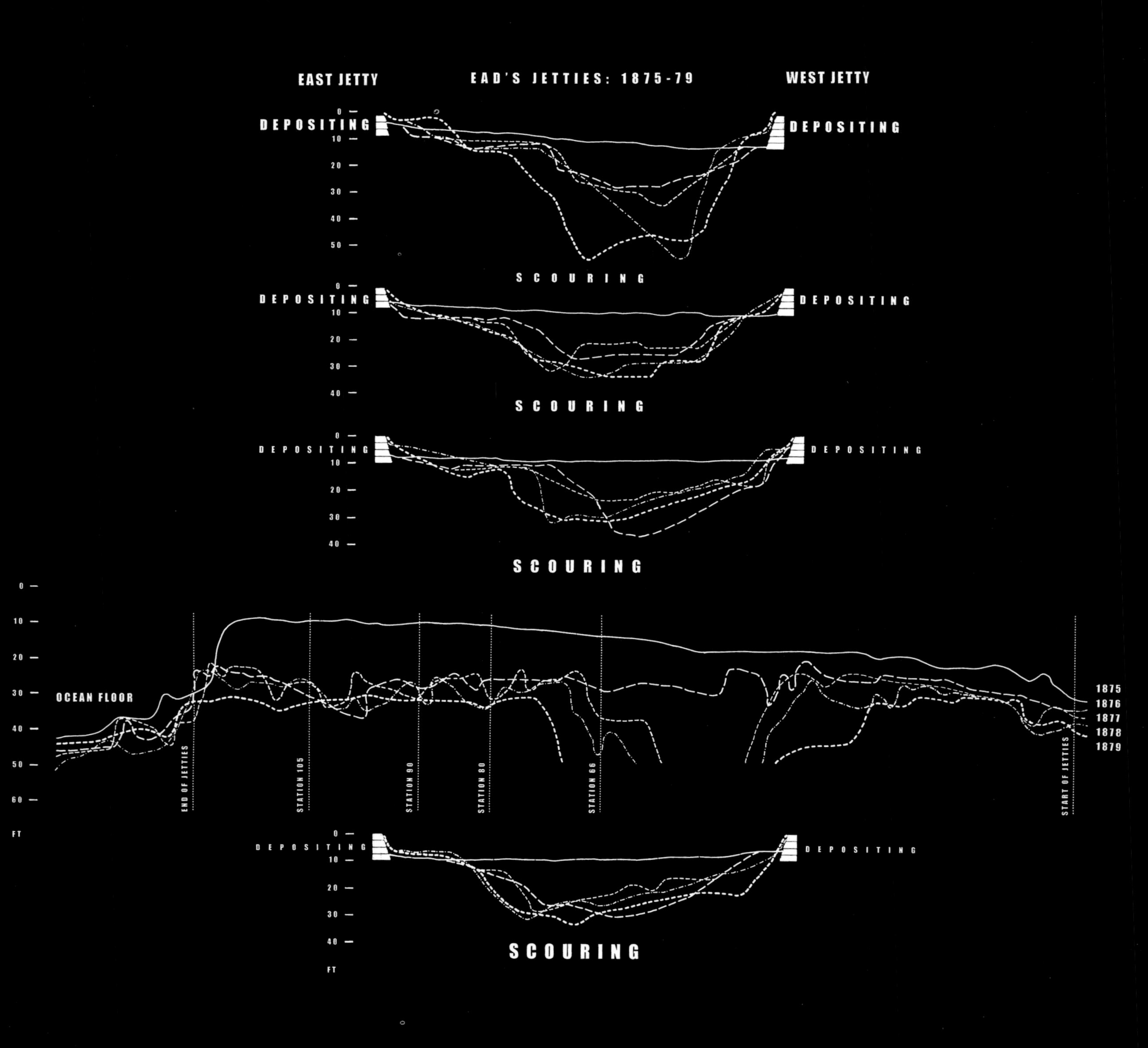

OPENING SOUTH PASS

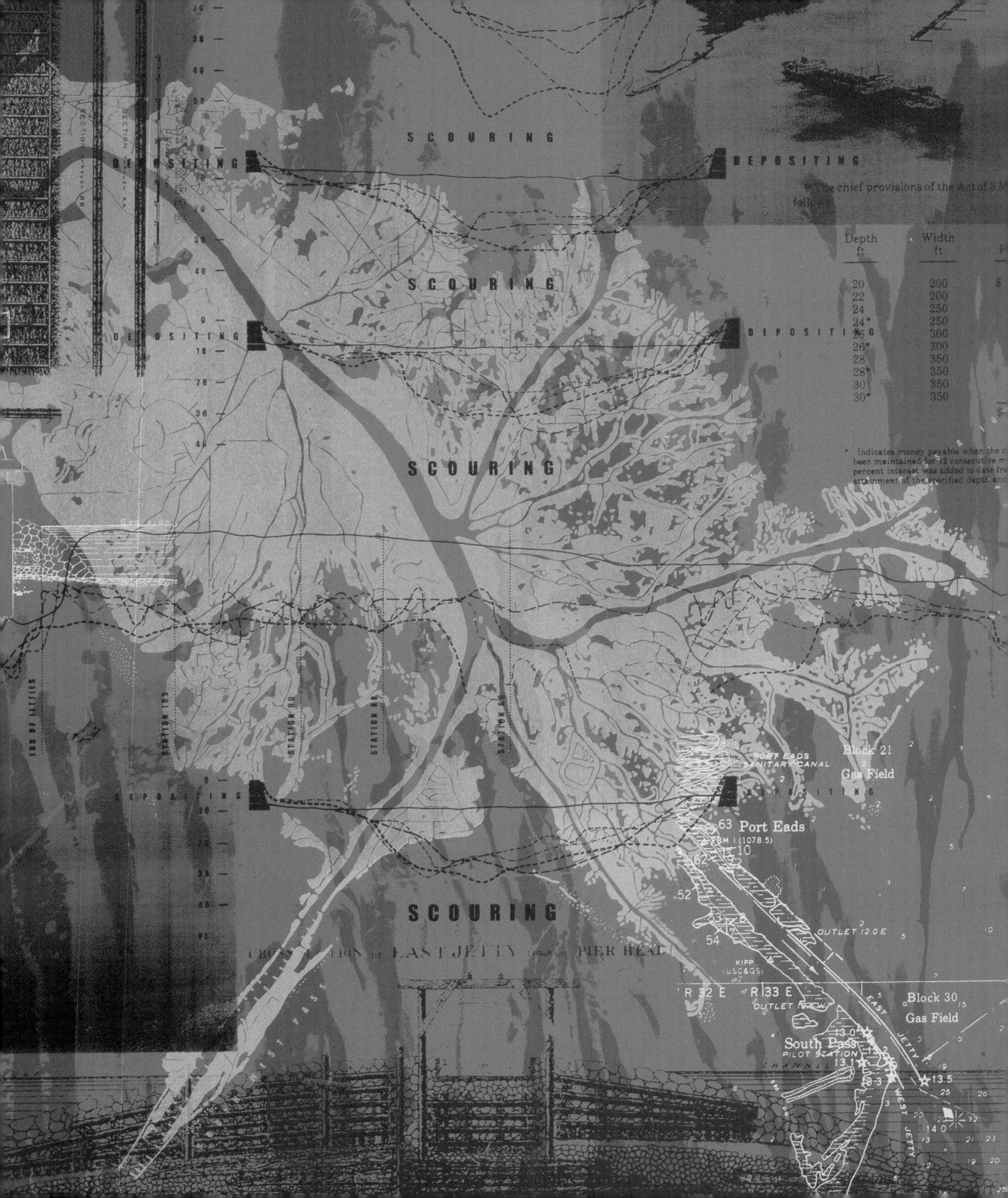

SCOURING
DEPOSITING
DEPOSITING
SCOURING
DEPOSITING
DEPOSITING
SCOURING
SCOURING
Depth ft
Width ft
20 200
22 200
24 250
24* 250
26 300
26* 300
28 350
28* 350
30 350
30* 350
STATION 90
STATION 66
END OF JETTIES
PORT EADS
SANITARY CANAL
Block 21
Gas Field
Port Eads
OUTLET 12.0 E
KIPP
(USC&GS)
R 32 E
R 33 E
Block 30
Gas Field
South Pass
PILOT STATION
EAST JETTY
WEST JETTY

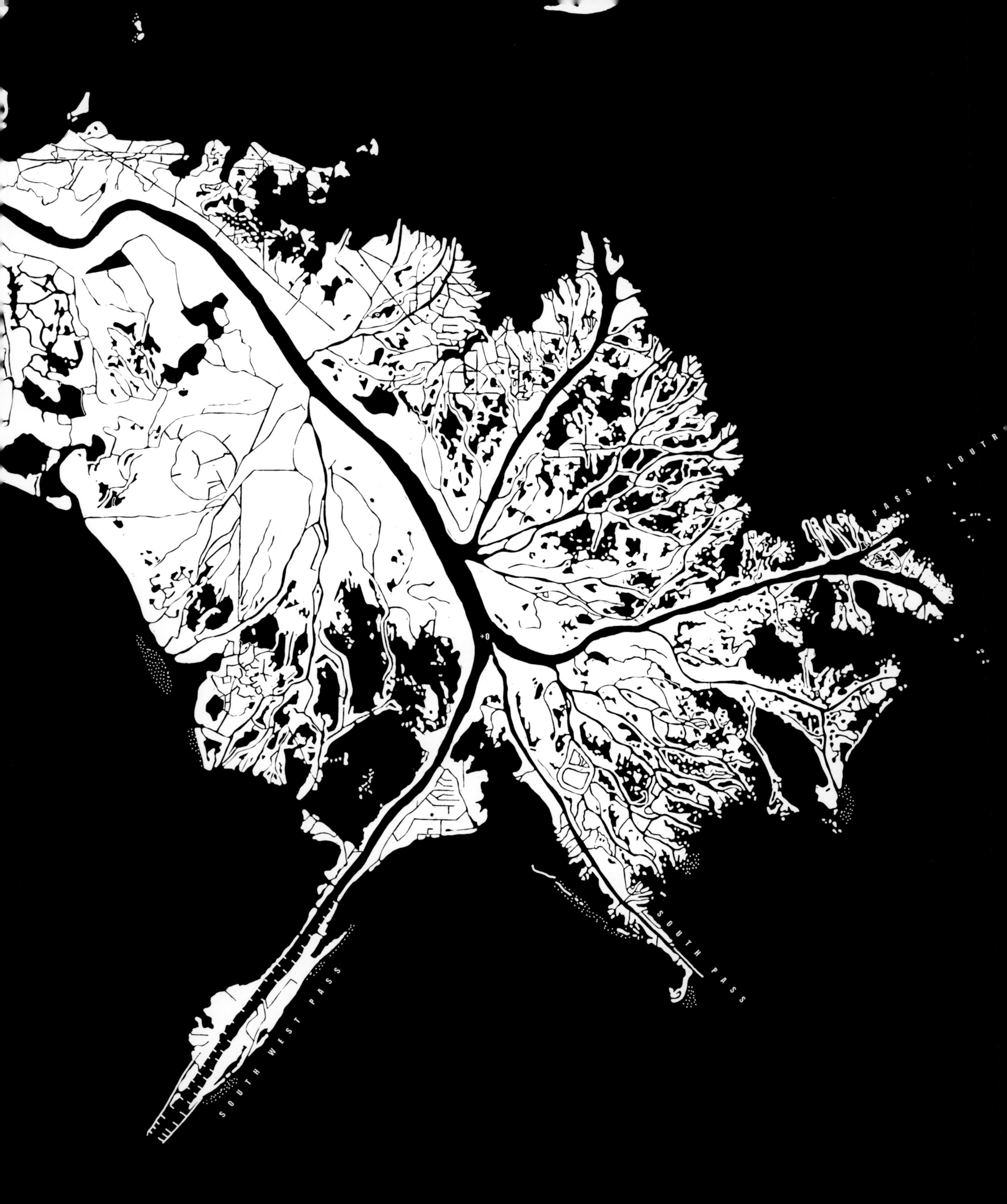
PASS A LOUTRE
SOUTH PASS
SOUTH WEST PASS

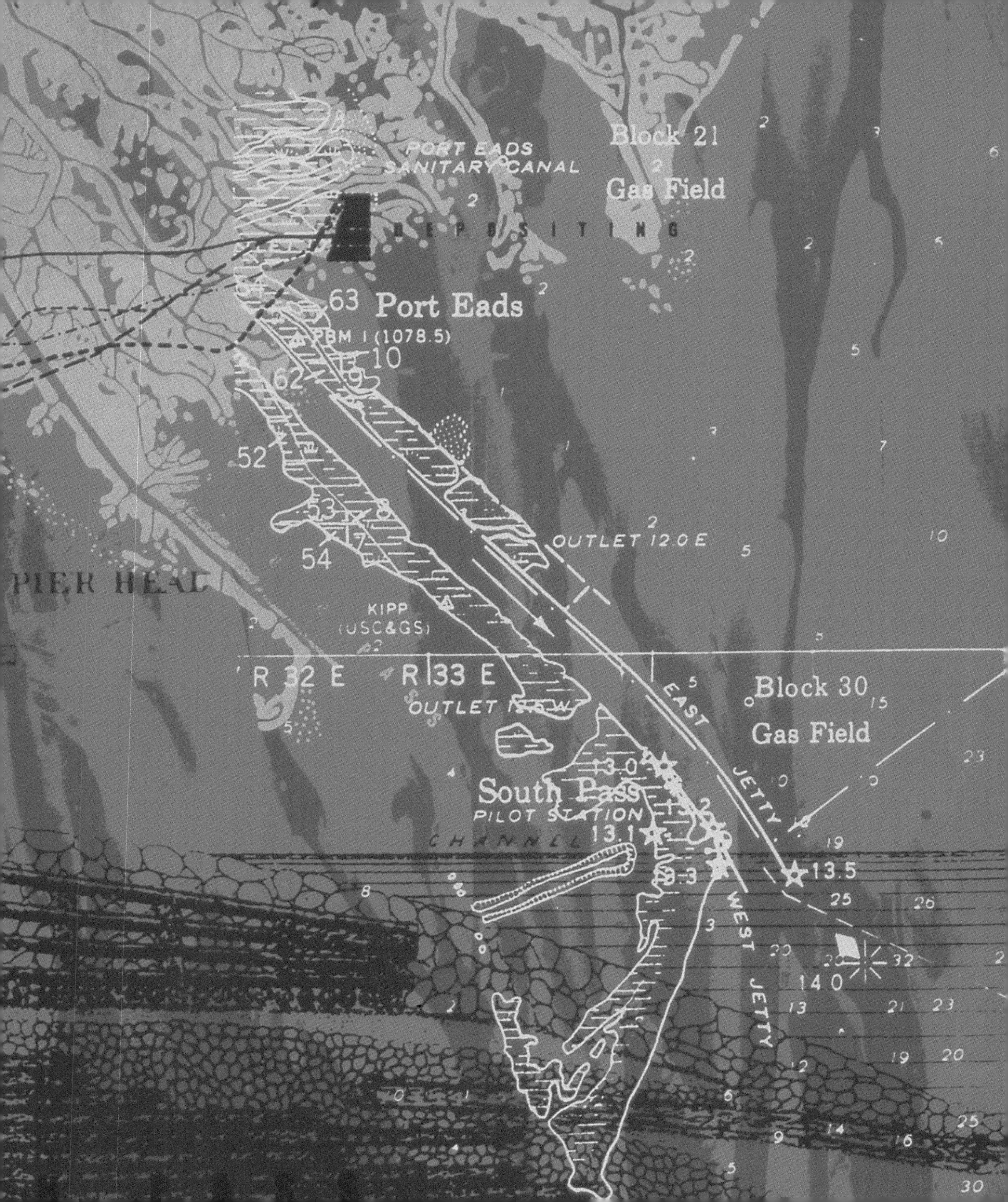

Block 21
Gas Field
PORT EADS
SANITARY CANAL
DEPOSITING
Port Eads
PBM I (1078.5)
OUTLET 12.0 E
KIPP
(USC&GS)
R 32 E
R 33 E
Block 30
Gas Field
EAST
JETTY
South Pass
PILOT STATION
CHANNEL
WEST
JETTY

ERODING CONTINENT

In a turn of events, land in the Mississippi Delta is diminishing. For centuries sediment coming out the mouth of the Mississippi extended the land not merely in this delta—the Plaquemine Complex, Balize Delta, or Birdfoot Delta—but in the river's four previous known deltas across the coast of Louisiana. Today it is as if the load is finally weighing down the continental shelf. Add to this rising sea levels and the extended confinement by levees and jetties that force the Mississippi to carry its load farther out into the deep waters of the Gulf, preventing it from making new land within the webs of the delta. Recent aerial photos show at least 30 percent less land than existed in maps made in the 1960s. But the load of the Mississippi—a bedload of primarily sand and a suspended load of clay and silt—is as visible today as it was to the Spanish explorers in 1686 who called it Cabo de Lodo (Cape of Mud).

The mud plume extending the mouth of the Mississippi River
[U.S. Army Corps of Engineers]

Writers over the centuries have presented a range of figures and pictures of the material washing off the continent. Mark Twain, working with the estimate of "able engineers"—406 million tons of mud annually—depicted "a mass a mile square and two hundred and forty-one feet high." Florence Dorsey visualized the "more than two hundred and seventy-six million cubic yards of material" per year as "a prism one mile square and two hundred and sixty-eight feet high." Willard Price described a "Big Sewer" that "pours into the sea about a cubic mile of mud a year." William Logan pictured twenty-five thousand railroad cars full of silt going into the Gulf daily. John Barry relays geologists' estimates of "an average of more than two million tons a day." Viewing this matter pouring out, which from these descriptions appears to be anywhere between 400 and 700 tons, and a quarter and 5 billion cubic yards a year, a dismayed traveler in 1827 wrote, "The first indication of our approach was the appearance of this mighty river pouring forth its muddy mass of waters, and mingling with the deep blue of the Mexican Gulf. I never beheld a scene so utterly desolate as this entrance of the Mississippi. Had Dante seen it, he might have drawn images of another Bolgia from its horrors." [4]

Tidal marshes at the mouth of the Mississippi
[U.S. Army Corps of Engineers]

To ecologists and resource prospectors, the entrance of the Mississippi is a haven rather than a Dantesque hell. It is an environment teeming with life, fragile and nonrenewable to one, abundant and sustainable to the other. The tidal marshes between the diverging flows of a continent hold in their saline waters and oyster grasses one of the richest spawning grounds for shrimp. Beneath those waters, as much as five miles below, is one of the largest oil repositories in America. Crisscrossing channels for pipelines and access routes to oil platforms, built out of concern more for a subterranean inert resource than for the living surface, have opened and erased large portions of the marsh.

Harvesting shrimp in the waters between river and sea [Louisiana Department of Wildlife and Fisheries]

Beyond the marshes in the distant marine waters that shrimp frequent in their life cycle are other sites visible only in maps—"ocean dumping," for example, in areas of water with depths of four hundred to six hundred feet. They demarcate sites for explosives and industrial waste. The delta that receives the spoils of a nation evidently extends far beyond the mud plume into "the deep blue of the Mexican Gulf."

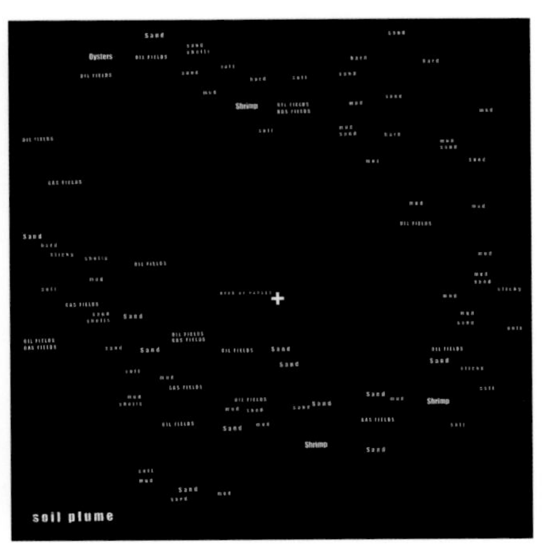

Eroding Continent
(screen print on paper, 44" x 30")

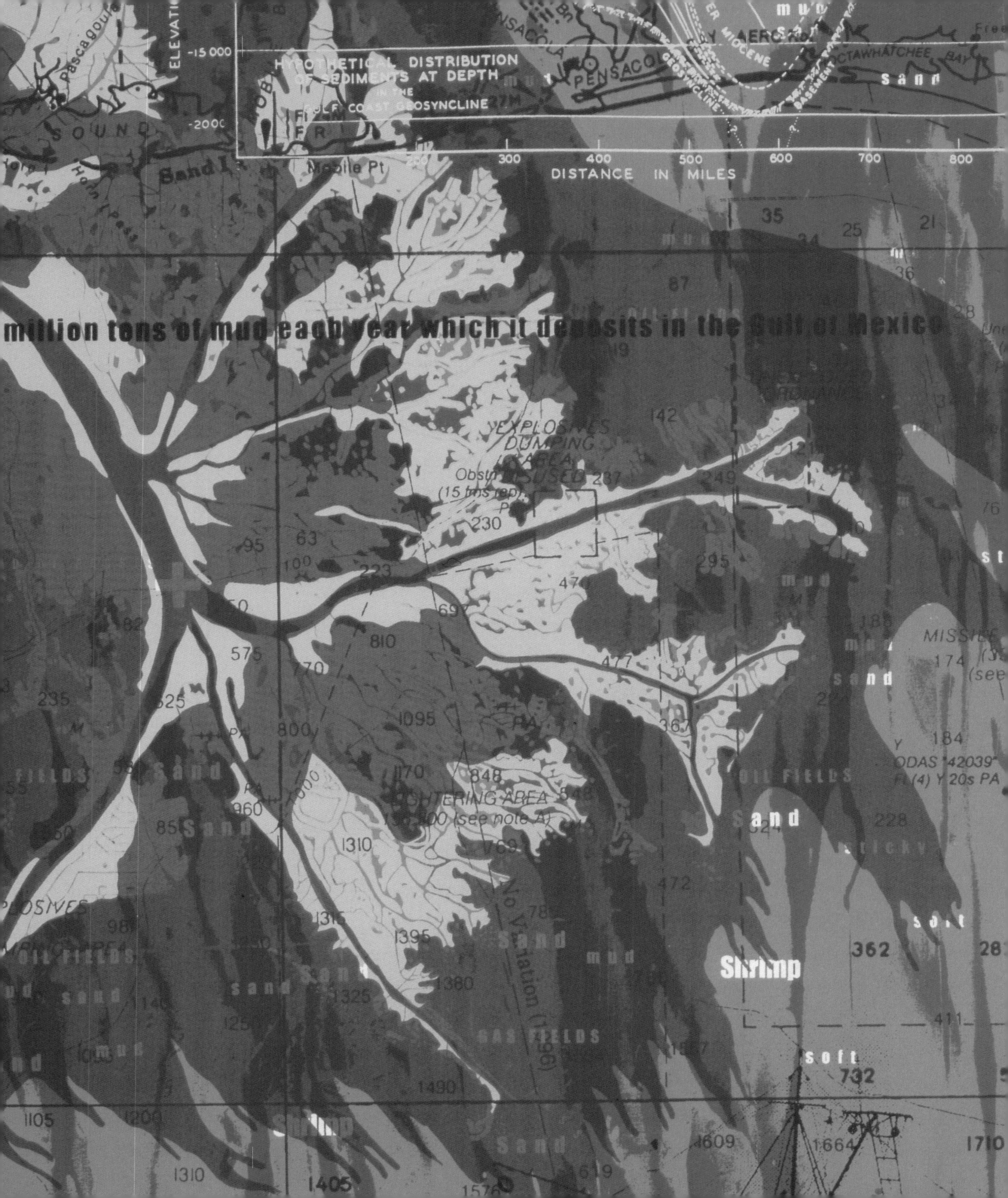

HYPOTHETICAL DISTRIBUTION OF SEDIMENTS AT DEPTH IN THE GULF COAST GEOSYNCLINE
ELEVATION
-15 000
DISTANCE IN MILES
300
400
500
600
700
800
Pascagoula
SOUND
Mobile Pt
PENSACOLA
CHOCTAWHATCHEE BAY
MIOCENE
GEOSYNCLINE
BASEMENT
million tons of mud each year which it deposits in the Gulf of Mexico
EXPLOSIVES DUMPING AREA DISUSED
(15 fms rep)
LIGHTERING AREA
(see note A)
No Variation (1996)
ODAS "42039"
Fl (4) Y 20s PA
OIL FIELDS
GAS FIELDS
sand
mud
soft
sticky
Shrimp

PILOTING EXCHANGE

It is one thing to see the mouth of the Mississippi as Land's End. The Corps of Engineers sees it this way.[5] It was undoubtedly seen as such by La Salle when he descended the length of the Mississippi in 1682, claiming the territory between the Great Lakes and the Gulf of Mexico for King Louis XIV of France. He was seeking the boundary of a continent and found it. It is perhaps also the view of riverboat pilots who journey the extensive waterways of the Mississippi system, their skills requiring the presence of land and unsuited to the open seas.

It is quite another thing, however, to see the delta as the Beginning of Land. Geologists see it this way. In the Mississippi Delta, they see the primeval forces that have made the alluvial valley from Cape Girardeau in the Commerce Hills of Missouri to the Gulf, as the Mississippi gradually silted the embayment left by falling sea levels beginning a million years ago. Seafaring explorers sailing on the trade winds in the fifteenth and sixteenth centuries must have seen the delta together with the rest of the vast Gulf rim this way as well, as they searched for an entrance to a continent.

The two views of the mouth are so different that La Salle, eager to affirm his claim for the king, could not find his way back in three years after emerging from it. Returning from Europe to the Gulf, he missed the delta by four hundred miles, baffled by the dry landscape where he landed. Some suggest his landing in Texas was a deliberate effort to extend French territorial claims as far west as possible, for cartographers at the time even showed the Mississippi as a branch of the Rio Grande.[6]

It is difficult to lose the entrance to the Mississippi or to extend territory today, but it still requires a complex exchange to negotiate the entry to the Mississippi navigational system, to shift perspective from the vast and deep expanse of ocean waters to the most channeled stream in the world only thirty-five feet deep at its entrance. The crucial point of exchange is Pilottown, a one-boardwalk settlement on stilts that is reached by water. It is located between the Head of the Passes and the town of Venice on the west bank where the levee that has run continuously for fifteen hundred miles ends.

Pilottown

A bar pilot from Pilottown takes over a vessel arriving from the Gulf at the mouth of the passes of the Mississippi Delta. It is the point where the Mississippi is spilling its tremendous amounts of sediment kept from settling in the passes by Eads Jetties. Once through the passes a river pilot takes over from the bar pilot for the stretch to New Orleans. Here another pilot takes over for the "parking" on the heavily trafficked River Road between New Orleans and Baton Rouge. The cargo from these ships can, with towboats, climb through locks to the upper reaches of the Red, Arkansas, Missouri, Mississippi, Illinois, Ohio, and Tennessee rivers, or reach ports in Texas and Florida, via the Intracoastal Canal built just inside the Gulf coast. Pilottown, called by some "an amphibious settlement," is the end and beginning of a commercial empire.[7]

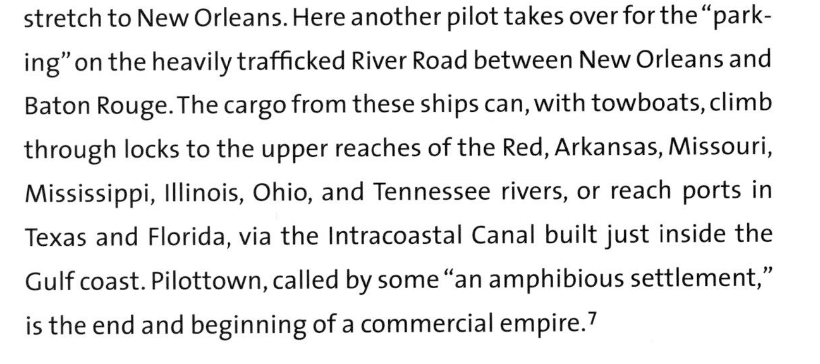

Pilot boats at the Head of the Passes
[U.S. Army Corps of Engineers]

Traffic jam at the entrance to New Orleans Harbor

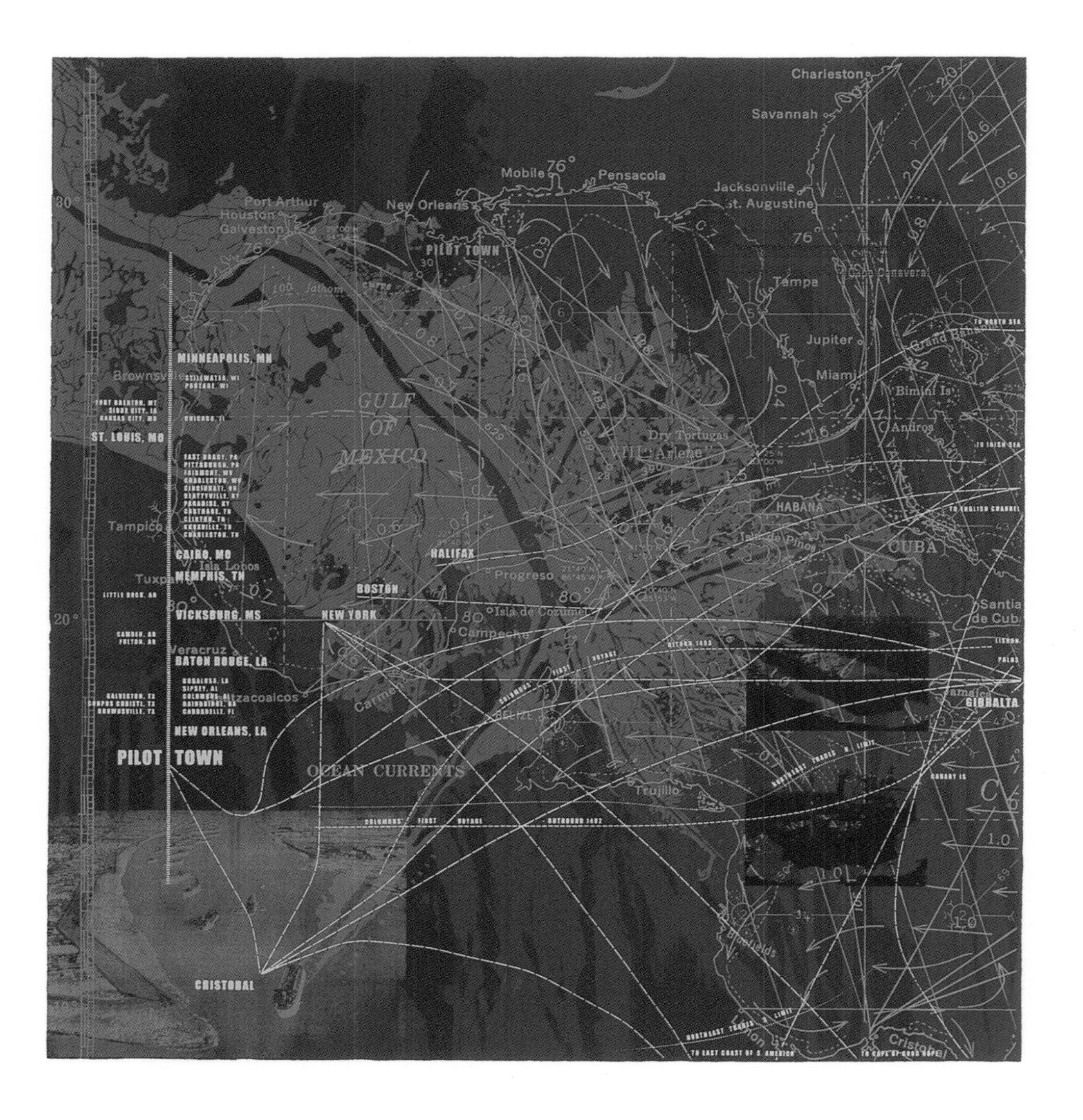

1890 LIGHT HOUSE AT SW PASS

Piloting Exchange
(screen print on paper, 44" x 30")

PILOT TOWN
100 fathom curve
MINNEAPOLIS, MN
STILLWATER, WI
PORTAGE, WI
GULF OF MEXICO
CHICAGO, IL
EAST BRADY, PA
PITTSBURGH, PA
FAIRMONT, WV
CHARLESTON, WV
CINCINNATI, OH
BEATTYVILLE, KY
PARADISE, KY
CARTHAGE, TN
CLINTON, TN
KNOXVILLE, TN
CHARLESTON, TN
CAIRO, MO
Isla Lobos
MEMPHIS, TN
VICKSBURG, MS
Veracruz
BATON ROUGE, LA
BOGALUSA, LA
SIPSEY, AL
COLUMBUS, AL
BAINBRIDGE, GA
CARRABELLE, FL
…tzacoalcos
NEW ORLEANS, LA
TOWN
OCEAN CURRENTS
Dry Tortugas
VIII "Arlene"
22°18'N 89°40'W
HALIFAX
21°50'N 85°03'W
Progreso
21°40'N 86°45'W
BOSTON
20°40'N 85°53'W
Isla de Cozumel
NEW YORK
Campeche
RETURN 1493
COLUMBUS' FIRST VOYAGE
Carmen
BELIZE
Trujillo
COLUMBUS' FIRST VOYAGE - OUTBOUND 1492

GATHERING BASIN

In New Orleans, the flow of the Mississippi keeps saline waters out of the city's drinking-water inlets while levees keep out the might of the Mississippi

The millions of people who inhabit the plains between the inner slopes of the Rockies and the Appalachians play a part in the flows of the Mississippi. Every shopping center, every drainage improvement, every square foot of new pavement in nearly half of the United States, John McPhee reminds us, accelerates runoff toward Louisiana. Not all of it, however, comes through the Birdfoot Delta. The flow through this millennium-old delta is restricted by the Corps of Engineers to 70 percent of the drainage of the Mississippi Basin. The remaining 30 percent is sent through the Atchafalaya, strongly favored by geologists to be the next delta complex, the sixth in as many millennia. It is a shift that the Corps, with its mandate to protect the people of the nation, wants to prevent at all costs, its interests visible in the massive control structures 315 miles upriver from the Head of the Passes.[8]

Emergent landscape at Land's End

It will require a catastrophic flood to defy the control structures at Simmesport, to unleash the agency that, as McPhee puts it, "has created most of Louisiana." But the Corps is prepared for such an event with Project Flood. At the time of the worst possible flood in the Mississippi basin, by the calculations of meteorologists, the engineers will dictate the measure and extent by which the Atchafalaya is favored. The Atchafalaya will get 50 percent and the Birdfoot Delta 42 percent, and 8 percent will be sent through the Bonnet Carré Spillway into Lake Pontchartrain. These divisions have been tested at a scale of 5.4 minutes to 24 hours on the Mississippi Basin Model, 360 miles north at Clinton outside of Jackson, Mississippi. It is in the image of this model that the Mississippi basin is worked today, its soils translated back from the coefficient of friction employed in the construction of that model, its flows worked like the waters passing through its pumps, valves, and pipes.[9]

In the delta is seen most vividly what John McPhee meant when he said that the Mississippi is the force that "created most of Louisiana." Those who live behind the levees undoubtedly catch a glimpse of this creative force in that moment of bewilderment when the Mississippi floods. There is human agency in this creative force today, as much agency as in the calculating, drawing, casting, laying, and testing of the Mississippi Basin Model. But between the river that it represents and that moment of bewilderment lies a vast, shifting, changing, and inexhaustible topography that Mark Twain, born in Hannibal on the Mississippi, called the "Body of the Nation."[10]

Abandoned pump and gauge of the Mississippi Basin Model at Clinton, Mississippi

The Birdfoot Delta
[U.S. Army Corps of Engineers]

Gathering Basin
(screen print on paper, 44" x 30")

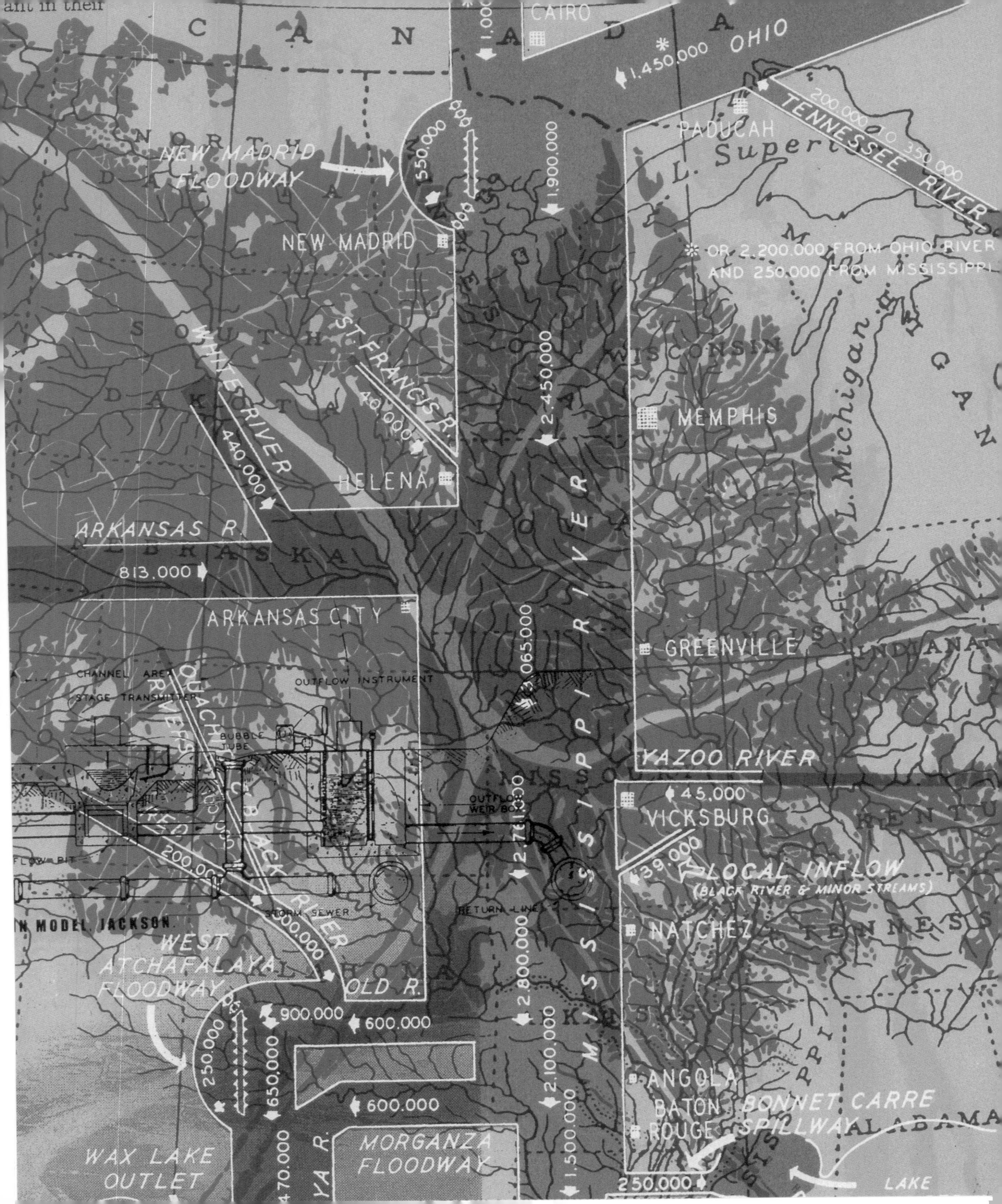
CANADA
CAIRO
1,000
1,450,000 OHIO
PADUCAH
200,000 TO 350,000
TENNESSEE RIVER
L. Superior
NEW MADRID FLOODWAY
550,000
1,900,000
NEW MADRID
* OR 2,200,000 FROM OHIO RIVER AND 250,000 FROM MISSISSIPPI
NORTH DAKOTA
SOUTH DAKOTA
WHITE RIVER
440,000
ST. FRANCIS R.
40,000
2,450,000
WISCONSIN
MEMPHIS
L. Michigan
MICHIGAN
HELENA
ARKANSAS R.
NEBRASKA
IOWA
813,000
ARKANSAS CITY
3,065,000
GREENVILLE
ILLINOIS
INDIANA
CHANNEL AREA
STAGE TRANSMITTER
OUTFLOW INSTRUMENT
BUBBLE TUBE
OUACHITA BLACK RIVERS
RED
MISSISSIPPI RIVER
YAZOO RIVER
MISSOURI
45,000
VICKSBURG
KENTUCKY
OUTFLOW WEIR BOX
2,781,000
39,000
LOCAL INFLOW
(BLACK RIVER & MINOR STREAMS)
200,000
FLOW PIT
STORM SEWER
RETURN LINE
BLACK RIVER
MODEL, JACKSON
NATCHEZ
TENNESSEE
WEST ATCHAFALAYA FLOODWAY
500,000
OKLAHOMA
OLD R.
2,800,000
900,000
600,000
KANSAS
250,000
650,000
2,100,000
ANGOLA
600,000
BATON ROUGE
BONNET CARRE SPILLWAY
ALABAMA
WAX LAKE OUTLET
470,000
MORGANZA FLOODWAY
1,500,000
250,000
LAKE

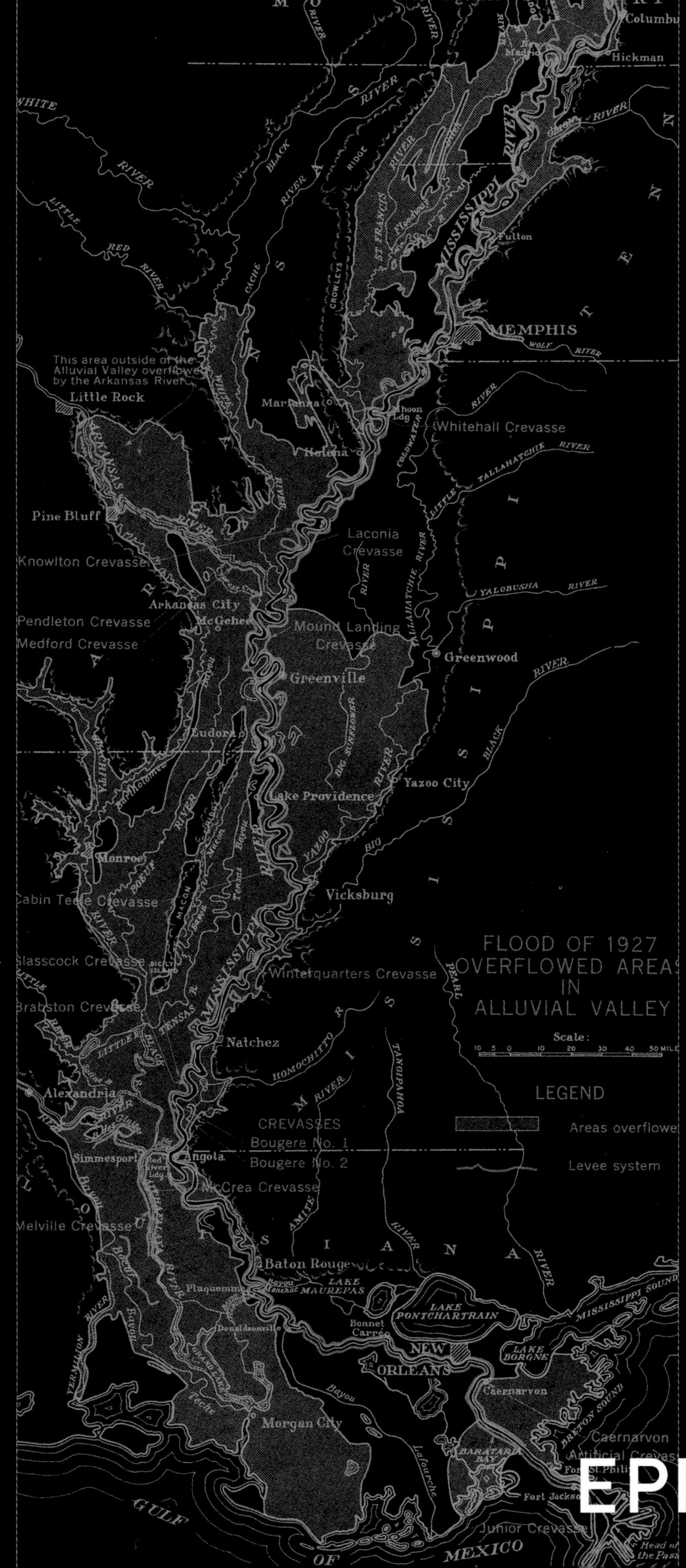

FLOOD OF 1927
OVERFLOWED AREAS
IN
ALLUVIAL VALLEY
Scale:
LEGEND
Areas overflowe
Levee system
CREVASSES
Bougere No. 1
Bougere No. 2
This area outside of the Alluvial Valley overflowed by the Arkansas River
Columbu
Hickman
MEMPHIS
Fulton
Little Rock
Pine Bluff
Marianna
Helena
Whitehall Crevasse
Laconia Crevasse
Knowlton Crevasse
Arkansas City
McGehee
Pendleton Crevasse
Medford Crevasse
Mound Landing Crevasse
Greenwood
Greenville
Eudora
Yazoo City
Lake Providence
Monroe
Vicksburg
Cabin Teele Crevasse
Glasscock Crevasse
Winterquarters Crevasse
Brabston Crevasse
Natchez
Alexandria
Simmesport
Angola
McCrea Crevasse
Melville Crevasse
Baton Rouge
Plaquemine
Donaldsonville
LAKE MAUREPAS
LAKE PONTCHARTRAIN
Bonnet Carre
NEW ORLEANS
LAKE BORGNE
Caernarvon
Caernarvon Artificial Crevasse
Morgan City
BARATARIA BAY
BRETON SOUND
MISSISSIPPI SOUND
Fort Jackson
Junior Crevasse
Head of the Passes
GULF OF MEXICO
MISSISSIPPI RIVER
MEMPHIS
WHITE RIVER
ARKANSAS RIVER
RED RIVER
YAZOO RIVER
PEARL RIVER

EPILOGUE

It does not take a flood to return the Lower Mississippi landscape to terra incognita. Behind the everyday working scene in a time of peace, this terrain is as open to speculative possibilities as it was when Ptolemy mapped the river that De Soto saw "overflow the meadows in an immense flood," or when La Salle "took possession of that river, of all rivers that enter it and of all the country watered by them." Our journey, beginning with the Mississippi Basin Model and continuing through meanders, flows, banks, and beds, offered glimpses of the unfathomable depth of this enigmatic landscape. We caught sight of it under the obvious divisions of river and settlement, soil and labor, silt and cargo, Atchafalaya and Mississippi, wealth and poverty; beyond the infrastructural acts of settlement—surveying land, building levees, draining swamps—that, given accounts of this once forbidding terrain, are truly magnificent feats of habitation; embedded in the everyday human practices that mark the rhythms of this emergent land, such as cultivating, dredging, towing, and crossing; and, finally, behind the images that capture this dynamic landscape for use by designers, including maps, paintings, diagrams, data sheets, construction drawings, photographs, texts, and conversations.

This flux that we sought to engage in our travels was far from the first or primordial nature that European pioneers, ignoring the Indians, saw: "a new creation starting from the deep," as William Darby said of the Atchafalaya Basin in 1812. It was rather a cultivated landscape, but one whose cultivation is part of the enigma. The Greenville cutoffs, like the bends they abandoned, the Delta blues, like the voices they released, the cottonfields, like the swamps they replaced, the dredging records, like the soil they chart, the levees, like the river they channel, are all cultivations that in the everydayness of their working, the articulation of their measure, and the emergence of their relations are very much part of the terra incognita that is the Lower Mississippi today.

Enjoyable and revelatory as our journey has been for us, we see it opening the Lower Mississippi to an imagination that tends to be underplayed by professionals. Identities that are mere assertions on the part of early surveyors and settlers are taken for granted as having a selfhood, a real and unquestionable presence in this landscape—river, settlement, water, soil. These identities are read with increasing sophistication and their limits, nature, meanings, and relations are subject to ongoing disputes and new agreements, as evidenced, for example, in the "movement" of Stack Island, the modifications of the Atchafalaya Floodway levels, the shift to Project Flood, and the increasingly green vocabulary of the Corps. But they are hardly questioned in themselves. Perhaps there is too much at stake in the war between river and settlement to allow playing with this firm ground; perhaps it is an unnecessary distraction to call the divisions of this landscape into question in a terrain facing the everyday threat of flood. But on the threshold of a fourth century of war based on asserted identities it is time to admit this first act of the settler into public debate, not for the sake of a critical society, but to allow for new imaginings of the Lower Mississippi.

We therefore see the recounting of our journey as the seed of a public project. This project is directed not toward resolving the problem of flood but toward keeping the possibility of reimagining the Lower Mississippi alive. Describing getting lost in Leningrad with a map in hand, E. F. Schumacher writes in *A Guide for the Perplexed:* "The maps of real knowledge, designed for real life, did not show anything except things that allegedly could be proved to exist. The first principle of the . . . map-makers seemed to be 'If in doubt, leave it out,' or put it into a museum. It occurred to me, however, that the question of what constitutes proof was a very subtle and difficult one. Would it not be wiser to turn the principle into its opposite and say 'If in doubt, show it prominently'? After all, matters that are beyond doubt are, in a sense, dead; they do not constitute a challenge to the living." In an age of "floodware"—computer programs that "make it easy" to navigate the data of flood—what new maps do we need to reveal that we cannot know and therefore must constantly challenge ourselves to negotiate the Lower Mississippi?[1]

Consumed by the polarities that rise with every flood (more control of the river versus withdrawal of settlement, humans versus nature, development versus environment), professionals pay little attention to what is really at stake in the Lower Mississippi. It is not just money, life, economy, or ecosystem, but the openness of imagination necessary to inhabiting an enigmatic landscape, a landscape very much in doubt not for lack of proof of what exists but because what constitutes proof is subtle and difficult. As the war between settlement and river continues, perpetuated as much by extremists as by mediators and moderates, other ways of seeing and inhabiting the Lower Mississippi will emerge. They will not begin with flood but with flux.

NOTES

PREFACE

1

See, for example, Stanley A. Changnon (ed.), *The Great Flood of 1993: Causes, Impacts, and Responses* (Boulder, Colo.: Westview, 1998).

INTRODUCTION

1

John McPhee, *The Control of Nature* (New York: Farrar, Straus & Giroux, 1989), 26.

2

William Alexander Percy, *Lanterns on the Levee* (New York: Knopf, 1941); Champ Clark, *Flood* (Alexandria, Va.: Time-Life, 1982), 65; Mark Twain, *Life on the Mississippi* (New York: Harper & Brothers, 1951), 225.

3

Willard Price, *The Amazing Mississippi* (London: Heinemann, 1962), 115.

4

Captain Frederick Marryat, *A Diary in America with Remarks on Its Institutions* (New York: D. Appleton, 1839), 124; Captain Frederick Marryat, *A Diary in America with Remarks on Its Institutions,* Part 2 (London: Longman, Orne Brown, Green & Longmans, 1839), 1:248–49.

5

John Francis McDermott, *Lost Panoramas of the Mississippi* (Chicago: University of Chicago Press, 1958), 85.

6

Captain Frederick Marryat, *A Diary in America with Remarks on Its Institutions,* Part 2 (London: Longman, Orne Brown, Green & Longmans, 1839), 1:247.

7

Captain Basil Hall, *Travels in North America in the Years 1827 & 1828* (Philadelphia: Carey, Lea & Carey, 1829), 2:281; Hon. Charles Augustus Murray, *Travels in North America in the Years 1834, 1835, & 1836* (London: Richard Bentley, 1839), 1:232–33.

8

Images in design, James Corner writes, "are neither mute nor neutral depictions of existing and projected conditions, of secondary significance to their object; rather they are active parts of the process, engendering, unfolding, and participating in emergent realities" ("Operational Eidetics: Forging New Landscapes," *Harvard Design Magazine*, fall 1998, 22).

9

See, for example, Denis Cosgrove (ed.), *Mappings* (London: Reaktion Books, 1999); James Duncan and David Ley (eds.), *Place/Culture/Representation* (New York: Routledge, 1993).

SITE 0 BASIN

1

U.S. Army Corps of Engineers, *The Mississippi Basin Model* (Vicksburg, Miss.: Waterways Experiment Station, n.d.) 10.

2

William H. Goetzmann and Glyndwr Williams, *The Atlas of North American Exploration* (New York: Prentice Hall, 1985), 65.

3

D. O. Elliott, *The Improvement of the Lower Mississippi River for Flood Control and Navigation* (Vicksburg, Miss.: U.S. Waterways Experiment Station, 1932), 1:5.

4

Elliott 1932, 3:141.

5

P. A. Feringa and W. Schweizer, *One Hundred Years Improvement on the Lower Mississippi River* (Vicksburg, Miss.: Mississippi River Commission, 1952), 7.

6

John M. Barry, *Rising Tide: The Great Mississippi Flood of 1927 and How It Changed America* (New York: Simon & Schuster, 1997), 41.

7

Ibid., 90.

8

Willard Glazier, *Down the Great River* (Philadelphia: Hubbard Brothers, 1888).

9

Elliott, 1932, 2:171; Albert Cowdrey, *Land's End* (Alexandria, Va.: U.S. Army Corps of Engineers, 1977), 37.

10

Major General Edgar Jadwin, Chief of Engineers, quoted in ibid., 44; J. P. Kemper, *Rebellious River* (Boston: Bruce Humphries, 1972), 109.

11

Ibid., 119.

12

U.S. Army Corps of Engineers, *The Mississippi Basin Model*, 3.

SITE 1 MEANDERS

1

Twain 1951, 153; Willard Price, *The Amazing Mississippi* (London: Heinemann, 1962), 135–36.

2

Barry 1997, 95.

3

Franklin Rosemont, Preface to Paul Garon, *Blues and the Poetic Spirit* (New York: Da Capo, 1978), 7.

4

Paul Oliver, *The Story of the Blues* (London: Design Yearbook, 1969), 19.

The attitude of preservation was pointed out by the blues musician Arthneice Jones, whom we met in Clarksdale, Miss., the Mecca of the Delta blues, in 1997.

Charlie Patton (born between 1881 and 1887) is referred to as the "Father of the Delta Blues," the "Voice of the Delta," and the "King of the Delta Blues." He grew up in the Delta. See Clyde Woods, *Development Arrested: The Blues and Plantation Power in the Mississippi Delta* (New York: Verso, 1998), 110.

5

Twain 1951, 156; Andrew Brookes, *Channelized Rivers: Perspectives for Environmental Management* (New York: John Wiley, 1988), 18.

6

Price 1962, 123.

7

Barry 1997, 97.

8

Tony Dunbar, *Delta Time* (New York: Pantheon, 1990), 11.

9

Twain, 1951, 3–4.

10

Hubert B. Herring, "An Island Keeps Rolling Along," *New York Times*, November 5, 1995.

11

Frederick Law Olmsted, *A Journey in the Seaboard Slave States* (New York: Negro Universities Press, 1968), 394.

12

Geologist quoted in Robert L. Brandfon, *Cotton Kingdom of the New South* (Cambridge: Harvard University Press, 1967), 29. On boll weevils, ibid., 123–24. On the pamphlet, Barry 1997, 109.

13

Dunbar 1990, 6.

14

Pete Daniel, *Deep'n as It Come: The 1927 Mississippi River Flood* (Fayetteville: University of Arkansas Press, 1996), 15.

15

New York Times, May 6, 1927.

16

Mississippi Valley Committee, *Report of the Mississippi Valley Committee of the Public Works Administration* (Washington, D.C.: United States Government Printing Office, 1934), 3; Daniel 1996, 3.

17

Barry 1997, 98, 97.

18

Dunbar 1990, 8.

19

Captain Frederick Marryat, *A Diary in America with Remarks on Its Institutions*, Part 2 (London: Longman, Orne Brown, Green & Longmans, 1839), 1:247; Dunbar 1990, 68.

SITE 2 FLOWS

1

McPhee 1989, 9–11. Monk Ptolemy, who accompanied De Soto's expedition in this vicinity in 1542, provided the earliest map of the junction of the Mississippi, Red, and Atchafalaya rivers, which he drew in 1576. It shows that the Atchafalaya, then as now, served as an outlet for the Mississippi. The Army Corps of Engineers seems reassured by this comparison (see Feringa and Schweizer 1952, 2).

2

Ibid., 10.

3

John McPhee suggests the notion of the Cajun Triangle. Speaking of the navigation lock near Simmesport, he observes, "The adjacent terrain is Cajun country, in a geographical sense the apex of the French Acadian world, which forms a triangle in southern Louisiana, with its base the Gulf Coast from the mouth of the Mississippi almost to Texas, its two sides converging up here near the lock—and including neither New Orleans nor Baton Rouge" (McPhee 1989, 3).

4

Ibid., 23.

5
New York Times, May 19, 1927.
6
McPhee 1989, 80.
7
Martin Reuss, *Designing the Bayous: The Control of Water in the Atchafalaya Basin, 1800–1995* (Alexandria, Va.: Office of History, U.S. Army Corps of Engineers, 1998), 17.
8
Barry Jean Ancelet, Jay D. Edwards, and Glen Pitre, *Cajun Country* (Jackson: University Press of Mississippi, 1991), xviii.
9
McPhee 1989, 31; Peter S. Feibleman, *The Bayous* (New York: Time-Life, 1973).
10
Reuss 1998, McPhee 1989.

SITE 3 BANKS

1
Mary Ann Sternberg, *Along the River Road: Past and Present on Louisiana's Historic Byway* (Baton Rouge: Louisiana State University Press, 1996), distinguishes the "River Road" from the "Great River Road." The latter refers to a route conceived by the Mississippi River Parkway Commission in 1938, running along the river from its source in Minnesota to the last point of solid land below New Orleans. It was intended to celebrate the history and culture of the Mississippi River Valley. River Road is a local name given to the roadways adjacent to the river levees. Sternberg extends it to include the settlements and culture that evolved along these roadways between Baton Rouge and New Orleans.
2
The "unwonted" and "fearfully wild" quotations were used by Frances Trollope to describe this landscape in her novel *The Life and Adventures of Jonathan Jefferson Whitlaw or Scenes on the Mississippi* (London: Richard Bentley, 1836). Almost every traveler who has left an account of this landscape has been struck by the huge effort required to make it habitable.
3
Cowdrey 1977, 1.
4
Barry 1997, 39.
5
Elliott 1932, 2:177–78.
6
Sternberg 1996, 29.
7
Twain 1951, 19.
8
Sternberg 1996, 48–49.
9
Twain 1951, 65, 81, 107.

SITE 4 BEDS

1
On the River of Palms, see Cowdrey 1977, 1. On Pánfilo de Narvaez, Goetzmann and Williams 1985, 33.
2
On the depth of the passes, E. L. Corthell, *A History of the Jetties at the Mouth of the Mississippi River* (New York: John Wiley, 1880), 13. On mud lumps, Roger T. Saucier, *Geomorphology and Quaternary Geologic History of the Lower Mississippi Valley* (Vicksburg, Miss.: U.S. Army Engineer Waterways Experiment Station, 1994), 1:150, and Corthell 1880, 16.

3

On tonnage shipped, Barry 1997, 89. Unlike levees, jetties were successful because they were in the stream, pushing the current all the time. Levees that are set back from the channel and only confine waters during flood do not have the same result.

4

Twain 1951, 2; Florence Dorsey, *Road to the Sea* (New York: Rinehart, 1947), 170; Price 1962, 179; Barry 1997, 39; William Bryant Logan, *Dirt: The Ecstatic Skin of the Earth* (New York: Berkeley, 1995), 117; Mrs. Trollope quoted in Twain 1951, 226.

5

Cowdrey 1977.

6

Goetzmann and Williams 1985, 88.

7

Price 1962, 178.

8

McPhee 1989, 46.

9

Ibid., 5.

10

Ibid.; Twain 1951, xv.

EPILOGUE

1

E. F. Schumacher, *Guide for the Perplexed* (London: Abacus, 1977), 11.

INDEX

Page numbers in *italic type* refer to illustrations